my **revision** notes

AQA A-level

BIOLOGY

Mike Boyle

HODDER EDUCATION
AN HACHETTE UK COMPANY

Hachette UK's policy is to use papers that are natural, renewable and recyclable products and made from wood grown in sustainable forests. The logging and manufacturing processes are expected to conform to the environmental regulations of the country of origin.

Orders: please contact Bookpoint Ltd, 130 Park Drive, Milton Park, Abingdon, Oxon OX14 4SE. Telephone: (44) 01235 827720. Fax: (44) 01235 400454. Email education@bookpoint.co.uk

Lines are open from 9 a.m. to 5 p.m., Monday to Saturday, with a 24-hour message answering service. You can also order through our website: www.hoddereducation.co.uk

ISBN: 978 1 4718 4219 1

First published in 2016 by

Hodder Education,
An Hachette UK Company
Carmelite House
50 Victoria Embankment
London EC4Y 0DZ
www.hoddereducation.co.uk

Impression number 10 9 8 7 6 5 4 3 2 1
Year 2020 2019 2018 2017 2016

Cover photo reproduced by permission of Sebastian Duda/Fotolia

Typeset in Bembo Std Regular, 11/13 pts. by Aptara, Inc.

Printed in Spain

A catalogue record for this title is available from the British Library.

Get the most from this book

Everyone has to decide his or her own revision strategy, but it is essential to review your work, learn it and test your understanding. These Revision Notes will help you to do that in a planned way, topic by topic. Use this book as the cornerstone of your revision and don't hesitate to write in it — personalise your notes and check your progress by ticking off each section as you revise.

Tick to track your progress 

Use the revision planner on pages 4, 5 and 6 to plan your revision, topic by topic. Tick each box when you have:

- revised and understood a topic
- tested yourself
- practised the exam questions and gone online to check your answers and complete the quick quizzes

You can also keep track of your revision by ticking off each topic heading in the book. You may find it helpful to add your own notes as you work through each topic.

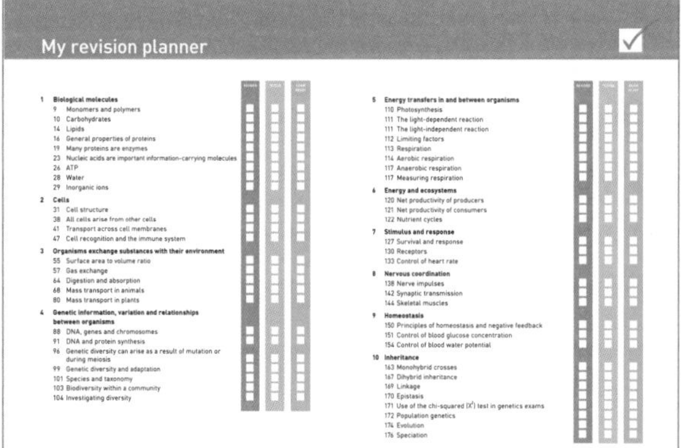

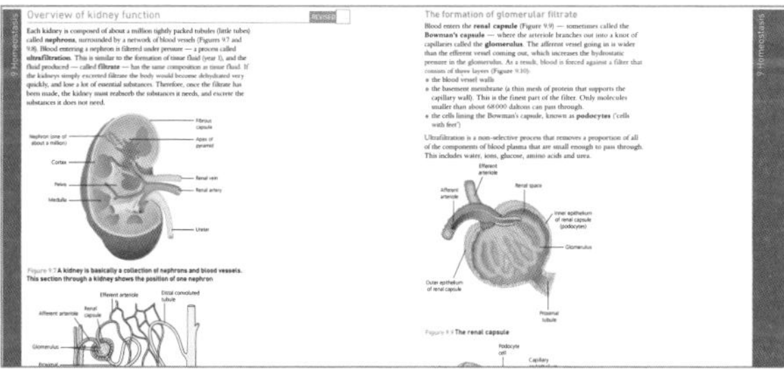

Features to help you succeed

Exam tips

Expert tips are given throughout the book to help you polish your exam technique in order to maximise your chances in the exam.

Typical mistakes

The author identifies the typical mistakes candidates make and explains how you can avoid them.

Now test yourself

These short, knowledge-based questions provide the first step in testing your learning. Answers are at the back of the book.

Definitions and key words

Clear, concise definitions of essential key terms are provided where they first appear.

Key words from the specification are highlighted in bold throughout the book.

Revision activities

These activities will help you to understand each topic in an interactive way.

Exam practice

Practice exam questions are provided for each topic. Use them to consolidate your revision and practise your exam skills.

Summaries

The summaries provide a quick-check bullet list for each topic.

Online

Go online to check your answers to the exam questions and try out the extra quick quizzes at **www.hoddereducation.co.uk/myrevisionnotes**

My revision planner

Exam practice answers and quick quizzes at www.hoddereducation.co.uk/myrevisionnotes

Countdown to my exams

6–8 weeks to go

- Start by looking at the specification — make sure you know exactly what material you need to revise and the style of the examination. Use the revision planner on pages 4, 5 and 6 to familiarise yourself with the topics.
- Organise your notes, making sure you have covered everything on the specification. The revision planner will help you to group your notes into topics.
- Work out a realistic revision plan that will allow you time for relaxation. Set aside days and times for all the subjects that you need to study, and stick to your timetable.
- Set yourself sensible targets. Break your revision down into focused sessions of around 40 minutes, divided by breaks. These Revision Notes organise the basic facts into short, memorable sections to make revising easier.

REVISED

2–6 weeks to go

- Read through the relevant sections of this book and refer to the exam tips, exam summaries, typical mistakes and key terms. Tick off the topics as you feel confident about them. Highlight those topics you find difficult and look at them again in detail.
- Test your understanding of each topic by working through the 'Now test yourself' questions in the book. Look up the answers at the back of the book.
- Make a note of any problem areas as you revise, and ask your teacher to go over these in class.
- Look at past papers. They are one of the best ways to revise and practise your exam skills. Write or prepare planned answers to the exam practice questions provided in this book. Check your answers online and try out the extra quick quizzes at **www.therevisionbutton.co.uk/myrevisionnotes**
- Use the revision activities to try out different revision methods. For example, you can make notes using mind maps, spider diagrams or flash cards.
- Track your progress using the revision planner and give yourself a reward when you have achieved your target.

REVISED

One week to go

- Try to fit in at least one more timed practice of an entire past paper and seek feedback from your teacher, comparing your work closely with the mark scheme.
- Check the revision planner to make sure you haven't missed out any topics. Brush up on any areas of difficulty by talking them over with a friend or getting help from your teacher.
- Attend any revision classes put on by your teacher. Remember, he or she is an expert at preparing people for examinations.

REVISED

The day before the examination

- Flick through these Revision Notes for useful reminders, for example the exam tips, exam summaries, typical mistakes and key terms.
- Check the time and place of your examination.
- Make sure you have everything you need — extra pens and pencils, tissues, a watch, bottled water, sweets.
- Allow some time to relax and have an early night to ensure you are fresh and alert for the examinations.

REVISED

My exams

A-level Biology Paper 1

Date:..

Time:..

Location:..

A-level Biology Paper 2

Date:..

Time:..

Location:..

A-level Biology Paper 3

Date:..

Time:..

Location:..

1 Biological molecules

Monomers and polymers

The biochemical basis of life

REVISED

All living things are made from just four basic types of **organic compound**: proteins, carbohydrates, nucleic acids and lipids.

These organic compounds perform similar functions in all organisms. For example, all organisms have DNA as their genetic material and all use it to make proteins according to the same code. The chemical reactions that occur inside cells are controlled by proteins called enzymes. The similarities and differences in these molecules provide clear evidence for evolution.

Some of these molecules are **polymers** — large molecules made from many repeated subunits called **monomers** joined in a chain. Monomers relate to polymers as follows:

- **Amino acids** join to make proteins.
- The **monosaccharide** glucose joins to make the polysaccharides starch, cellulose and glycogen.
- **Nucleotides** join to make the nucleic acids DNA and RNA.

Lipids, on the other hand, are not polymers.

An **organic compound** is one in which the molecules are based on carbon, i.e. proteins, carbohydrates, nucleic acids and lipids.

A **polymer** is a long chain of repeated units. The individual units are called monomers.

A **monomer** is one of the small similar molecules that join together to form a polymer. Three particularly important monomers are: amino acids, glucose and nucleic acids.

Hydrolysis and condensation

REVISED

Think of these reactions as 'breaking down' and 'building up again'. Large molecules are broken down into smaller ones by **hydrolysis**. Small molecules are built up into larger ones by **condensation**.

The food we eat contains a lot of polymers. Digestion involves breaking down these large molecules so that they are simple, soluble and can be absorbed into the blood (Figure 1.1). Large molecules are hydrolysed by enzymes to produce smaller molecules. Once inside the body, smaller molecules are built up into large ones by condensation. For example, we might hydrolyse the protein in a piece of chicken into amino acids that can be absorbed into the body. These amino acids could be used to build up the proteins the body needs such as **haemoglobin**, enzymes and muscle protein.

ABCDEFGH	→	A + B + C + D + E + F + G + H	→	GFEADBCH
Large molecule	Hydrolysis	Smaller molecules	Condensation	Large molecule

Figure 1.1 Digestion involves hydrolysis and condensation

Exam tip

Condensation and hydrolysis are common themes for exam questions. Make sure you get them the right way round. Condensation reactions *produce* water, whereas hydrolysis reactions *use* water. These reactions are the exact opposite of each other.

A **hydrolysis** reaction breaks a chemical bond between two molecules and involves the use of a water molecule.

A **condensation** reaction joins two molecules together with the formation of a chemical bond and involves the elimination of a molecule of water.

Haemoglobin is the red pigment that transports oxygen.

Now test yourself

TESTED

1 What type of reaction will join monomers to form polymers?
2 Which type of reaction will break polymers down into monomers?

Answers on p. 203

Carbohydrates

Monosaccharides and disaccharides

REVISED

Carbohydrates contain just three elements: carbon, hydrogen and oxygen. Simple carbohydrates are called sugars, which are all sweet, soluble compounds whose names end in *–ose*. There are two types of sugar: **monosaccharides** (single sugars) and **disaccharides** (double sugars).

The three monosaccharides you need to know are glucose, galactose and fructose. The three disaccharides you need to know are maltose, sucrose and lactose.

Exam tip

Try not to get confused with G words. Glycogen is a polysaccharide. Glucose and galactose are monosaccharides (simple sugars).

Glucose

Glucose is a vital molecule and you need to be able to draw its structure (Figure 1.2). The other two monosaccharides, **galactose** and **fructose**, are isomers of glucose — they have the same atoms but in a slightly different arrangement.

Glucose is our main energy source: our cells respire glucose most of the time. This means that the energy in the glucose is released and used to make ATP. In turn, ATP can provide instant energy for activities such as the contraction of muscles (see p. 24).

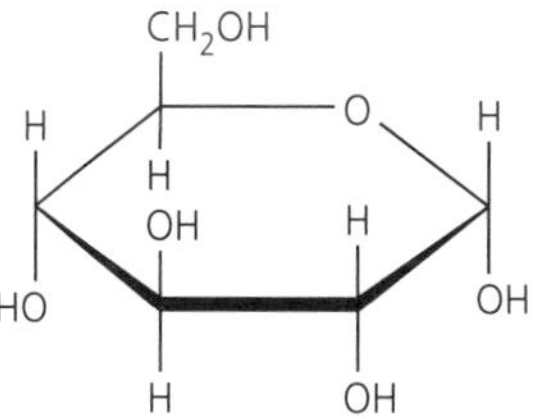

Figure 1.2 The structure of an (alpha) glucose molecule

α-glucose and β-glucose

Glucose has two **isomers**, α-glucose and β-glucose (Figure 1.3). There is a slight but important difference between these molecules. α-glucose polymerises to form the energy storage compounds starch and glycogen, whereas β-glucose polymerises into cellulose, a compound with completely different properties.

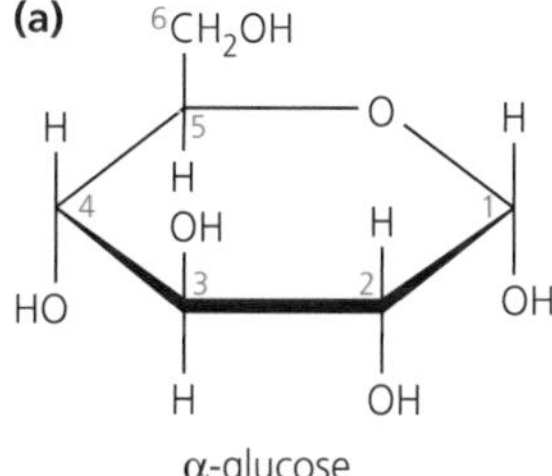

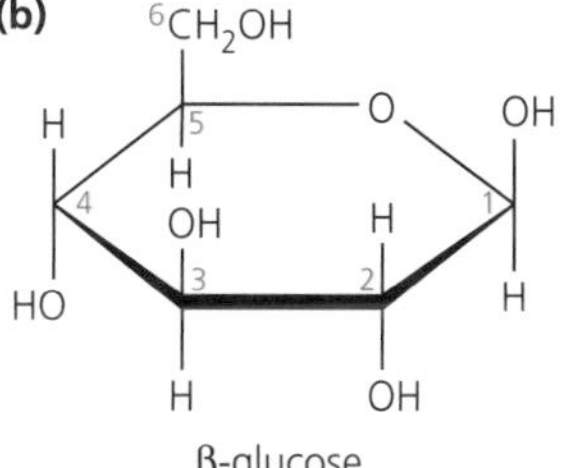

Figure 1.3 Comparing molecules of (a) α-glucose and (b) β-glucose. Note that the carbon atoms are numbered clockwise from the oxygen. The key difference is the position of the –OH group on carbon 1.

A **condensation reaction** between two α-glucose molecules forms a **glycosidic bond** that produces **maltose** (Figure 1.4). Starch is formed by simply repeating the process.

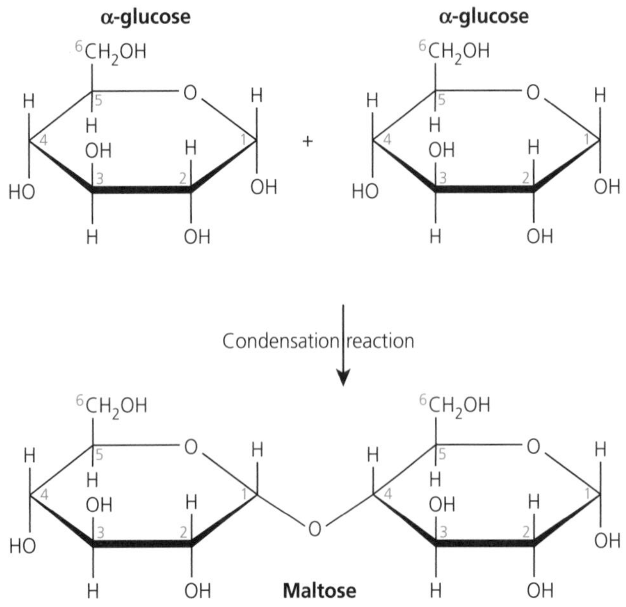

Figure 1.4 Two α-glucose molecules join to form a molecule of the disaccharide maltose. The bond formed is called a glycosidic bond and it is based around a shared oxygen atom

Exam tip

You only need to know the structure of glucose, not galactose or fructose.

Sucrose and lactose

These two other disaccharides are made from different monosaccharides:

- one molecule of **sucrose** (cane sugar) is formed by the condensation of one α-glucose molecule and one fructose molecule
- one molecule of **lactose** (milk sugar) is formed by the condensation of one α-glucose molecule and one galactose molecule

Exam tip

For the exam, make sure you can draw four key molecules: glucose, glycerol, a fatty acid and an amino acid.

Typical mistake

When drawing OH groups, the oxygen is always attached to the carbon and the hydrogen is always attached on the outside of the oxygen. So, when drawing molecules it should always be –OH or HO–, not –HO or OH–.

Polysaccharides

REVISED

Polysaccharides are also called complex carbohydrates. The three polysaccharides you need to know are starch, glycogen and cellulose. These are large molecules made from hundreds or thousands of glucose molecules joined together. Their large size makes them insoluble.

Starch

Starch is the main energy storage compound in plants. Plants make glucose by photosynthesis and then convert it into starch for storage, so it does not take up too much space and does not make the water potential of the cytoplasm too low. Starch is not one compound but a mixture of two: amylose and amylopectin (Figure 1.5):

- **Amylose** is a straight chain polymer of glucose, which means that it is one long spiral molecule with just two ends — there is no branching.
- **Amylopectin**, in contrast, is branched so that there are many more ends. This is important because new glucose units can only be added or released from the ends. Amylopectin can therefore be built up and broken down much more quickly than amylose.

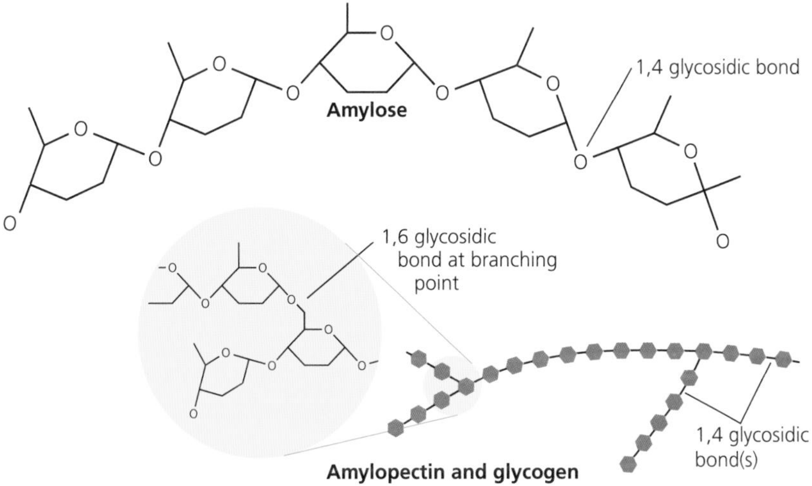

Figure 1.5 The structure of amylose and amylopectin

A storage compound needs to be compact, insoluble and available when needed. Starch does this role perfectly because:

- the spiral shape of amylose makes it compact
- its huge size makes it insoluble
- the branching of amylopectin provides lots of ends that can release glucose quickly

Glycogen

Learning about **glycogen** is easy: it has the same structure as amylopectin, but it is more frequently branched. As a result, it can be broken down and built up more quickly, which matches the greater energy needs of animals.

- Most of our glycogen is stored in the liver and muscles.
- When blood glucose levels are low, glycogen is broken down to restore levels.
- Think of glycogen as short-term energy storage, whereas lipids are long term.

Cellulose

Cellulose is the most common polysaccharide in the world. After water, it accounts for a large percentage of the weight of most plants, including trees. It is a key component of wood.

Starch and glycogen are polymers of α-glucose, whereas cellulose is a polymer of β-glucose. Cellulose has one key property: it forms fibres that have great strength. β-glucose molecules join together in the same way as α-glucose: by condensation reactions that form glycosidic bonds. The difference is that β-glucose units condense to form long, straight unbranched chains. When they lie alongside each other, in parallel, many hydrogen bonds form along the whole length, so that long, strong fibres form. Many microfibrils become glued together by a mixture of shorter glucose chains to form a tough wall that resists expansion (Figure 1.6).

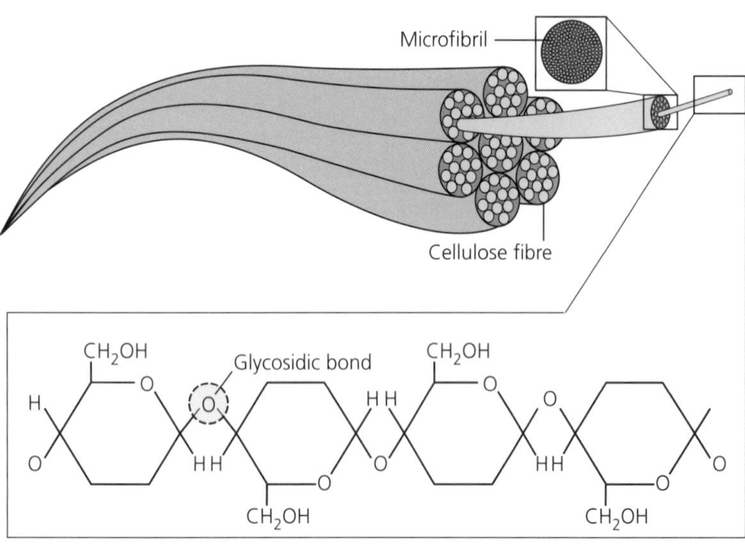

Figure 1.6 **In cellulose, chains of β-glucose molecules lie parallel to form strong microfibrils**

Now test yourself

TESTED

3 Explain how the structure of cellulose is related to its function.

Answer on p. 203

Testing for reducing and non-reducing sugars

REVISED

Benedict's solution is used to test for the presence of **reducing sugars** (e.g. glucose). It can also be used, indirectly, to test for **non-reducing sugars** (e.g. sucrose). For reducing sugars, dissolve the test substance in water, add Benedict's reagent and heat to almost boiling point. In a positive result, an orange precipitate forms. If it is negative, it stays blue. For non-reducing sugars, a Benedict's test will be negative. So, add a few drops of dilute hydrochloric acid and boil. Then neutralise the solution by adding a few drops of dilute sodium hydroxide. Finally, add Benedict's reagent and reheat again. If positive, a precipitate forms and the solution changes colour from blue to orange. This happens because the non-reducing sugar has been split into its constituent reducing sugars.

The Benedict's test can also be quantitative — it can be used to determine *how much* reducing sugar is present. There are several ways of doing this, in increasing order of accuracy:

1. Assessing the depth of colour by eye (it goes from vaguely greenish to deep orange).
2. Assessing the depth of colour using a colorimeter.
3. By filtering, drying and weighing the orange precipitate (which is copper oxide).

Now test yourself

TESTED

4 What is the difference between a qualitative test and a quantitative test?

Answer on p. 203

Testing for starch

REVISED

Testing for the presence of starch is simple — add a few drops of **iodine solution**. In a positive result, the sample changes colour from yellowish brown to deep blue/black.

Lipids

Lipids are what most people think of as fats and oils. The two vital properties of lipids are:

- they do not mix with water
- they store a lot of energy compared with an equivalent amount of carbohydrate or protein

The two types of lipid you need to know about are triglycerides and phospholipids. Overall, triglycerides store energy, whereas phospholipids make plasma membranes.

Triglycerides

REVISED

Triglycerides are formed by the **condensation** of one molecule of **glycerol** attached to three molecules of **fatty acid** (Figure 1.7). Triglycerides are the main energy storage compounds in animals. They contain the elements carbon, hydrogen and oxygen.

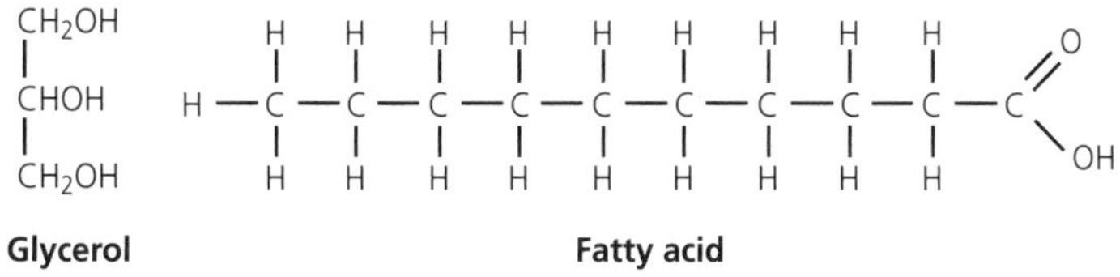

Figure 1.7 The basic structure of glycerol and a fatty acid

Fatty acids are also called organic acids or carboxylic acids (Figure 1.8). They all have a –COOH group attached to a carbon chain. The carbon chain is known as the R group, giving fatty acids the general formula RCOOH. Glycerol is always the same, but different fatty acids produce different triglycerides (Figure 1.9).

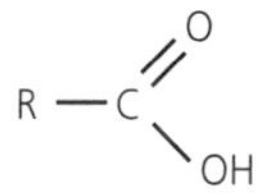

Figure 1.8 The generalised structure of a fatty acid. The hydrocarbon chain is just replaced by an R

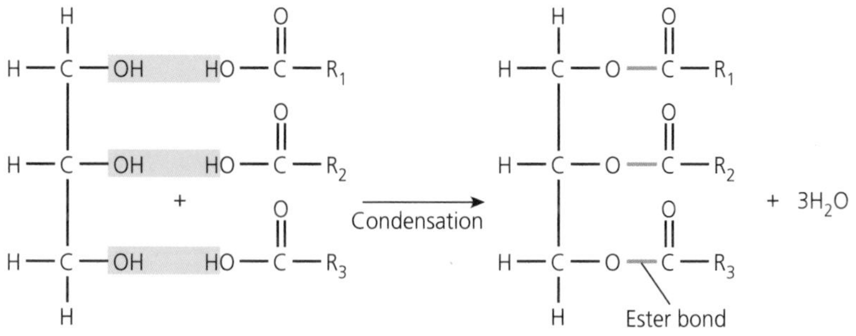

Figure 1.9 A triglyceride molecule consists of four subunits: one glycerol joins by condensation to three fatty acids. The new bonds are called ester bonds

Fatty acids

The R group of fatty acids can vary in two ways:

- the number of carbon atoms in the chain
- the number of C=C bonds in the chain

If there are no C=C bonds, the fatty acid is said to be **saturated** because it has as much hydrogen as possible. If there is one C=C bond, there must be two fewer C–H bonds and the fatty acid is said to be **unsaturated**. If there are two or more C=C bonds, the fatty acid is **polyunsaturated** (Figure 1.10).

```
(a)  H   H   H   H   H   H   H   H   H   H   H   H   H   H   H   H   H      O
     |   |   |   |   |   |   |   |   |   |   |   |   |   |   |   |   |    //
 H — C — C — C — C — C — C — C — C — C — C — C — C — C — C — C — C — C — C
     |   |   |   |   |   |   |   |   |   |   |   |   |   |   |   |   |    \
     H   H   H   H   H   H   H   H   H   H   H   H   H   H   H   H   H     OH

(b)  H   H   H   H   H   H   H   H   H   H   H   H   H   H   H   H   H      O
     |   |   |   |   |   |   |   |   |   |   |   |   |   |   |   |   |    //
 H — C — C — C — C — C — C — C — C — C = C — C — C — C — C — C — C — C — C
     |   |   |   |   |   |   |   |           |   |   |   |   |   |   |    \
     H   H   H   H   H   H   H   H           H   H   H   H   H   H   H     OH

(c)  H   H   H   H   H   H   H   H   H   H   H   H   H   H   H   H   H      O
     |   |   |   |   |   |   |   |   |   |   |   |   |   |   |   |   |    //
 H — C — C — C — C — C — C = C — C — C = C — C — C — C — C — C — C — C — C
     |   |   |   |   |           |           |   |   |   |   |   |   |    \
     H   H   H   H   H           H           H   H   H   H   H   H   H     OH
```

Figure 1.10 Fatty acids: (a) saturated, (b) unsaturated and (c) polyunsaturated

> **Typical mistake**
>
> Candidates often make the mistake of stating that saturated fats have no double bonds. However, they do have a C=O bond, just no C=C bonds.

Fatty acids in animal triglycerides tend to be saturated and solid at room temperature (for example, lard). Plant fats tend to be unsaturated and are usually oils at room temperature.

Phospholipids

REVISED

Phospholipids differ from triglycerides in that they contain a **phosphate-containing group** (PO_4^-) instead of one of the fatty acid molecules. Phospholipids therefore contain the elements carbon, hydrogen, oxygen and phosphorus.

The phosphate-containing group is vital. Instead of the whole molecule repelling water, the phosphate group attracts water. The result is that you get a molecule with two ends: a phosphate group that attracts water and two fatty acid tails that repel water (Figure 1.11).

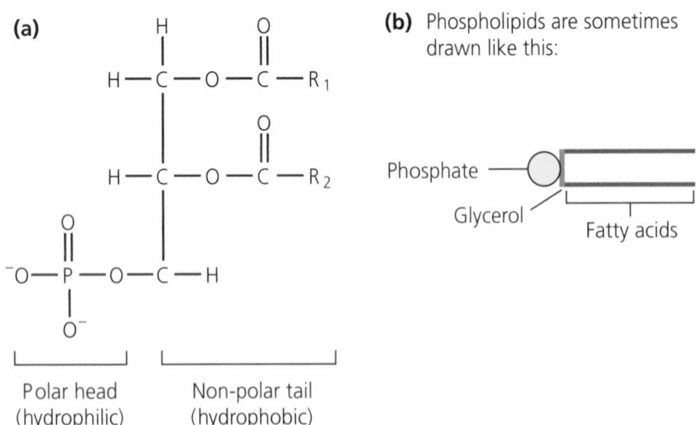

Figure 1.11 (a) A phospholipid. The head is hydrophilic (attracted to water) and the tails are hydrophobic (repelled by water). (b) A simplified phospholipid

In water, phospholipids automatically arrange themselves into a double layer — the basis of membranes in all cells (Figure 1.12). See p. 39 for more about membranes.

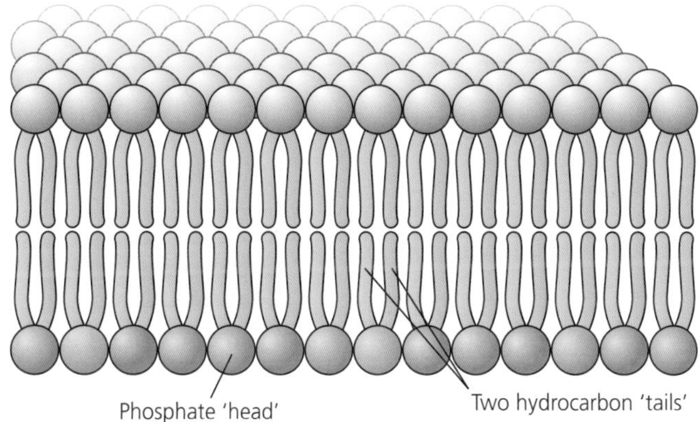

Figure 1.12 **Phospholipids form double layers when surrounded by water**

Testing for lipids

REVISED

The **emulsion test** is used to test for lipids. First, dissolve the test substance in ethanol, then filter. Then add water to the filtrate. If the test is positive, the substance will turn cloudy/milky because an emulsion has been formed. An emulsion is a suspension of lipid/alcohol droplets in water.

Now test yourself

TESTED

5 Describe how you would test a sample of cake for the presence of lipids.

Answer on p. 203

General properties of proteins

Proteins such as enzymes, antibodies and hormones are vital compounds in all living organisms. Their functions include holding your body together; allowing you to move, digest and absorb your food; controlling your body chemistry; and fighting off disease. Proteins are large, complex molecules that always contain the elements carbon, hydrogen, oxygen and nitrogen.

Amino acids

REVISED

Amino acids are the **monomers** from which proteins are made. Most amino acids have names that end in *–ine* such as proline, valine and serine. All of them have the same basic formula, which you may be asked to draw (Figure 1.13).

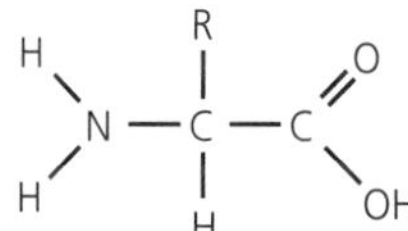

Figure 1.13 **A basic amino acid**

As you might expect from the name, there is an amino part and an acid part:

- the amino group is written as $–NH_2$
- the acid group is written as –COOH

It is the R group that is different. There are 20 different amino acids, so there are 20 different R groups. For example, in glycine the R group is simply a hydrogen atom, whereas in alanine the R group is CH_3.

Amino acids join by **condensation reactions**, forming **peptide bonds** in the process. There are two amino acids in a **dipeptide** (Figure 1.14), three in a tripeptide and many in a **polypeptide**. A **protein** is made from one or more polypeptides. A haemoglobin molecule, for example, is made from four polypeptides.

A **peptide bond** is the bond between two amino acids.

A **dipeptide** is formed when two amino acids are joined together by a peptide bond.

A **polypeptide** is a chain of many amino acids joined together by peptide bonds.

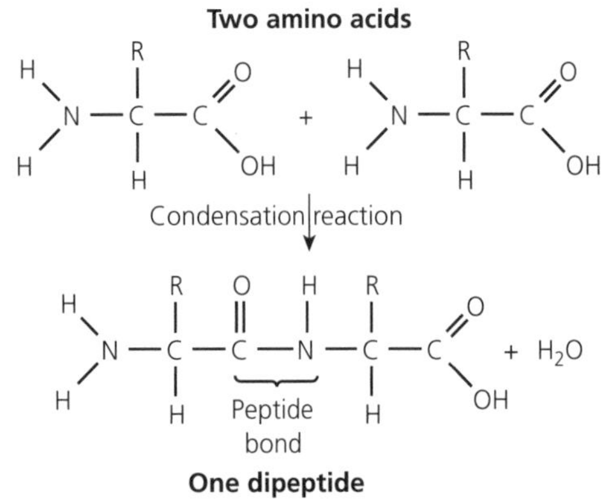

Figure 1.14 How a dipeptide is formed

Exam tip

You don't need to remember the names of the different amino acids, but you should learn their general structure.

Exam tip

A dipeptide has the backbone N–C–C–N–C–C. You can use this to check that you have drawn it correctly. Longer peptides all have the same backbone.

Exam tip

R does not stand for an element. Think of R as the **r**est of the molecule.

Revision activity

Draw a molecule of the amino acid alanine.

There is an infinite variety of proteins. Amino acids can join in any order; some may be repeated and others left out completely. Having 20 different monomers is the key to their variety. If there are 20 different amino acids, there must be 400 (20^2) different dipeptides and 8000 different tripeptides. So there can be a different enzyme for every reaction, a different antibody for every disease etc.

Revision activity

Using a whiteboard or blank piece of paper, draw two glucose molecules side by side. Highlight the atoms that will be involved in a condensation reaction and then draw the resulting disaccharide. Repeat this activity for two amino acids.

Protein structure

REVISED

Proteins are large, complex molecules so we divide the study of their structure into four levels (Figure 1.15).

1. **Primary structure** is the sequence of amino acids, such as valine – serine – proline – valine. Insulin, for example, is a small protein consisting of 51 amino acids. Many proteins have hundreds.
2. **Secondary structure** is the shape formed when part of the chain of amino acids becomes folded and coiled. The two most common secondary structures are an α-helix (a spiral shape) and a β-sheet (pleated like a folded sheet of paper).
3. **Tertiary structure** is the overall shape of the amino acid chain, i.e. the whole polypeptide. When in water, a wide variety of forces combine to twist, fold and bend the polypeptide into its most stable shape. There are many different areas of secondary structure within the tertiary structure.

4 **Quaternary structure** occurs when the protein has more than one polypeptide chain. If a protein is made from only one polypeptide chain, it does not have a quaternary structure.

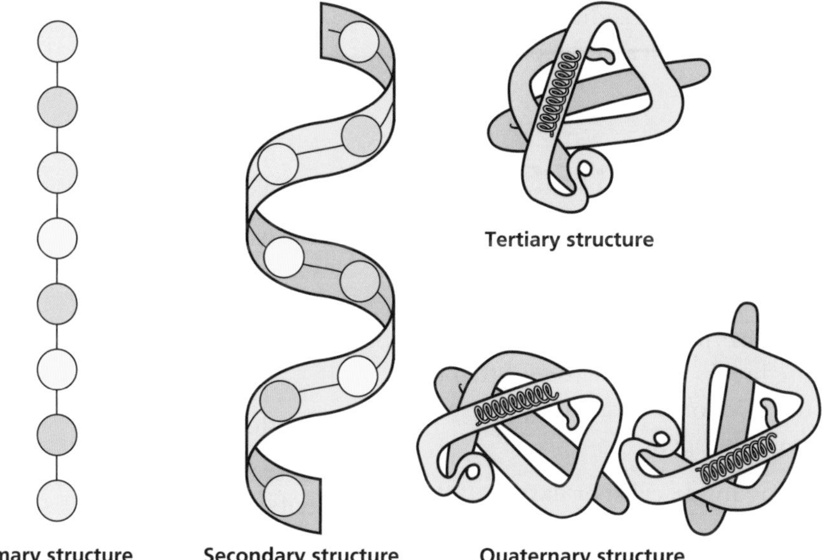

Figure 1.15 The four levels of protein structure

Protein function

REVISED

Proteins can be split into two main groups: **globular** or **fibrous**.

Globular proteins are generally rounded in shape. They tend to have a chemical function. Examples include:

- enzymes (see p. 17)
- membrane proteins that control the movement of ions and other substances in and out of the cell (see p. 39)
- hormones such as insulin (some hormones are proteins, others are not)
- haemoglobin (see p. 66)
- antibodies to help fight disease (see p. 45)

Fibrous proteins are usually long, thin molecules that generally have a structural function. Examples of fibrous proteins include:

- collagen, which gives strength to tissues such as bone, tendon and ligaments
- keratin, which gives strength to skin, hair and nails
- actin and myosin, which make muscle contract

Generally, the shape of globular proteins is maintained by relatively weak forces such as **hydrogen bonds** and **ionic bonds**. In contrast, big tough fibrous proteins rely more on strong **disulfide bridges**.

Now test yourself

TESTED

6 Which elements are contained in the following?
 (a) carbohydrates
 (b) lipids
 (c) proteins

7 Which of these are polymers?
 proteins triglycerides glucose sucrose starch glycogen phospholipids

Answers on p. 203

Testing for proteins

REVISED

The **biuret test** is a biochemical test used for detecting the presence of peptide bonds. First dissolve the test substance in water. Then add the biuret reagent (a mixture of copper sulfate and sodium hydroxide). A change in colour from blue to lilac/purple indicates the presence of protein.

Many proteins are enzymes

Enzymes are central to living things. Cells can be thought of as tiny units of controlled chemical reactions and it is the enzymes, together with hormones, that do the controlling. The reactions of the body are given the general name **metabolism**. Enzymes are named by adding *–ase* to the name of the substrate, so lactase breaks down lactose. Often the full name of the enzyme describes the reaction it catalyses. For example, glycogen synthetase makes (synthesises) glycogen.

Enzymes may be **intracellular** (working inside cells) or **extracellular** (working outside cells).

Intracellular proteins work inside cells.

Extracellular proteins work outside cells.

Tertiary structure

REVISED

Enzymes are globular proteins. They have a precise shape — their **tertiary structure**. The substance an enzyme works on is called the **substrate**. There is a part of the enzyme's surface called the **active site** into which the substrate fits. The active site and the substrate are **complementary** — the substrate fits into the active site, partly because it is the right shape and partly because the chemical charges match. The **specificity** of an enzyme refers to its ability to catalyse just one reaction or type of reaction. Only one particular substrate molecule will fit into the active site of the enzyme molecule.

The **tertiary structure** is the overall shape of the enzyme. Globular proteins are proteins which consist of polypeptide chains that are folded so that the molecule is roughly spherical and has a compact overall shape.

Enzymes as catalysts

REVISED

Enzymes are catalysts, so they speed up chemical reactions without being altered themselves. For any reaction to take place, the reactant molecules must collide and achieve the **transition state** — one in which the existing bonds are strained. After this, the existing bonds will break and new ones will form as new products are created.

Enzymes speed up reactions because they **lower the activation energy** needed to achieve the transition state (Figure 1.16). This is achieved by enabling the reaction to take a slightly different pathway by forming an **enzyme–substrate complex**. This complex alters the bonding in the substrate, enabling the bonds to be broken more easily.

Exam tip

In exam questions about enzymes, examiners expect to see you using the correct language, so use terms such as *tertiary structure*, *successful collisions*, *active site*, *enzyme–substrate complex*, *complementary* and *denatured*.

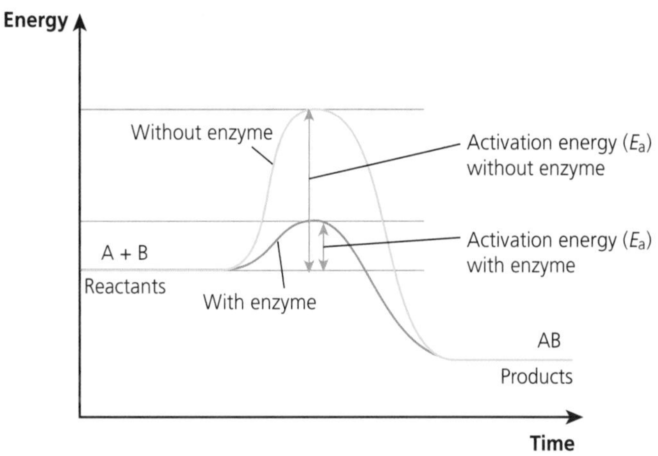

Figure 1.16 The effect of enzymes on activation energy

Now test yourself

TESTED

8 Explain how enzymes are able to speed up chemical reactions.
9 On Figure 1.16, label the point at which the transition state is achieved.

Answers on p. 203

The lock and key and induced-fit models

REVISED

The lock and key model states that the active site and the substrate are an exact match — they fit together perfectly (Figure 1.17). The active site is the lock and it only fits one key — the substrate. They come together to form an enzyme–substrate complex. Once formed, the reaction can be completed.

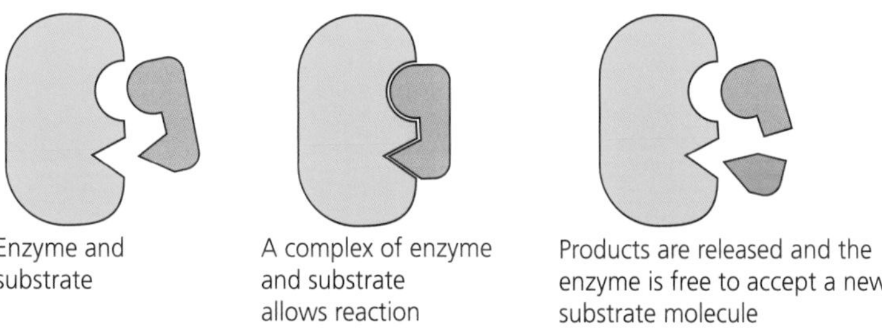

Figure 1.17 The lock and key model of enzyme action

The **induced-fit model** is a hypothesis that modifies the lock and key model. It helps to explain how enzymes are specific to their substrate.

The **induced-fit model** is slightly different from the lock and key hypothesis. The active site does not exactly match the substrate, but it alters its shape slightly when the substrate attaches. It is as if the active site moulds itself around the substrate (Figure 1.18).

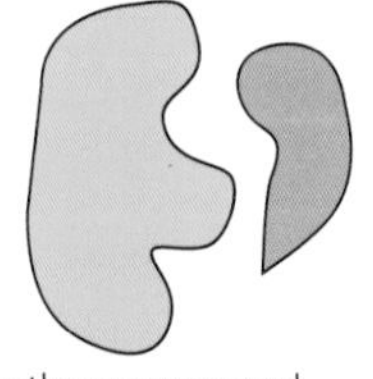

As the enzyme and substrate bind, a change of shape occurs

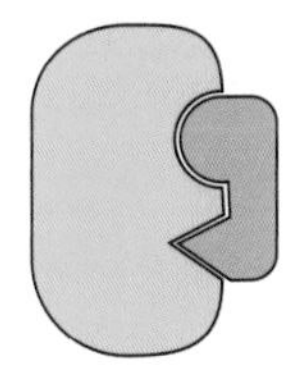

The reaction proceeds as the enzyme and substrate bind

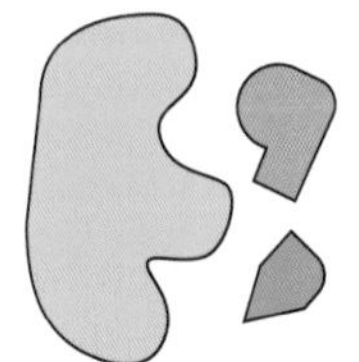

Products are released and the enzyme returns to its original shape

Figure 1.18 The induced-fit model of enzyme action

Most scientists now accept that the induced-fit model is a more accurate model of what really happens. The lock and key hypothesis suggests that the active site is rigid, but we know from the action of non-competitive inhibitors that the shape of active site can change. Techniques such as computer modelling have further supported the idea that the active site is flexible and that the catalysis is brought about by a complex interaction of the substrate and the amino acids at the active site.

Typical mistake

The active site and the substrate are *not* the same shape. They are complementary — one fits into the other.

Now test yourself

TESTED

10 Usually, one enzyme catalyses just one reaction. Explain why enzymes are so specific.
11 Explain how the lock and key model of enzyme action differs from the induced-fit model.

Answers on p. 203

The effects of surrounding conditions on enzyme activity

REVISED

Enzyme concentration

The higher the **enzyme concentration**, the greater the rate of reaction until substrate becomes the limiting factor. In living cells, however, enzymes exist in active and inactive forms. An enzyme is converted into its active form — effectively increasing its concentration — when the product it makes is needed. When the product is no longer needed, the enzyme is deactivated, effectively lowering its concentration again. It is a really clever self-regulating mechanism.

Substrate concentration

In a similar way to enzyme concentration, Figure 1.19 shows that the more substrate there is, the faster the enzyme will work until all the active sites are full all the time. After that, no matter what the **substrate concentration**, the reaction will not go any faster. All enzymes have a maximum rate at which they can work.

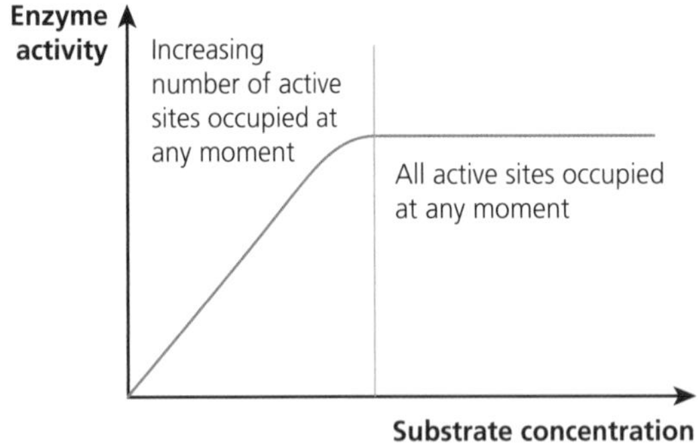

Figure 1.19 The effect of substrate concentration

Competitive and non-competitive inhibitors

Inhibitors are substances that slow down or stop enzyme action:

- **Competitive inhibitors** (Figure 1.20) are similar in shape to the substrate. They fit into the active site but cannot be converted into the product, so they simply get in the way. The more competitive inhibitors there are, the less chance there is of a successful collision between enzyme and substrate.

- **Non-competitive inhibitors** (Figure 1.21) bind to the enzyme away from the active site, but they alter the shape of the enzyme so that the substrate cannot fit into the active site. Effectively, non-competitive inhibitors switch off enzymes. If the inhibitor is removed, the enzyme functions as normal.

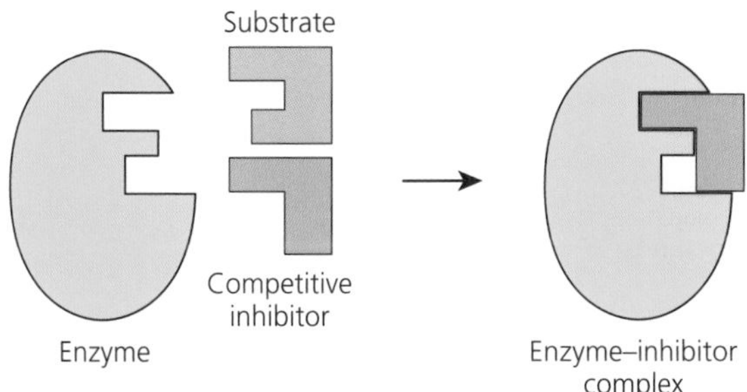

Figure 1.20 A competitive inhibitor competes with substrate molecules for the active site of the enzyme

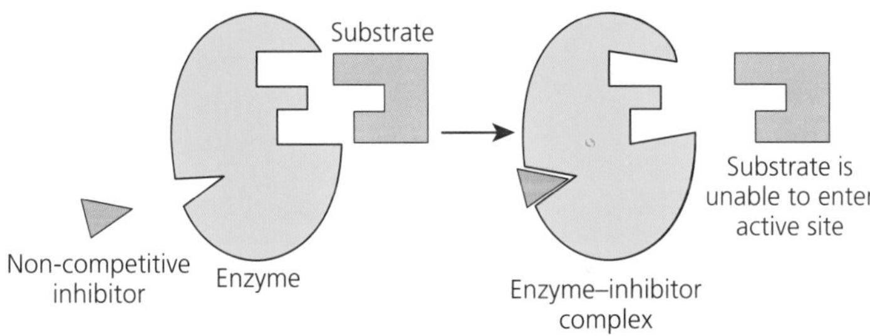

Figure 1.21 Non-competitive inhibitors bind to the enzyme somewhere other than the active site. This causes the active site to change shape

The key difference between the two types of inhibitor is that the effects of competitive inhibitors can be overcome by adding more substrate. It is simply a matter of improving the chances of a collision between enzyme and substrate compared with enzyme and inhibitor. With non-competitive inhibitors, on the other hand, it does not matter how much substrate is added — nothing can fit into the active site so there is no activity.

pH

The **pH** scale is a measure of acidity. On a scale of 1 to 14, pH 7 is neutral. A pH lower than 7 is acidic — there are a lot of H^+ ions. A pH of more than 7 is basic — there are more OH^- ions.

Every enzyme has an optimum pH. Most enzymes work inside cells and their optimum pH is around neutral. Some enzymes in the digestive system are exceptions; the stomach enzymes work best at around pH 2, whereas those that work in the small intestine have their optimum at about pH 8.

At optimum pH, the positive and negative charges on the active site and substrate are complementary. Away from this pH, the H^+ or OH^- ions neutralise the charges so that enzyme and substrate are no longer complementary. In addition, extremes of pH can denature enzymes.

Exam tip

If the hydrogen ion concentration of a solution is known, the pH can be calculated by substituting the hydrogen ion concentration into this formula:

$$pH = -\log_{10}[H^+]$$

Temperature

Figure 1.22 shows the relationship between enzymes and **temperature**. As the temperature increases, the rate of reaction increases because all the molecules bounce around faster and there is more chance of a collision between the enzyme and the substrate. Therefore, more enzyme–substrate complexes form, resulting in a product being formed more quickly.

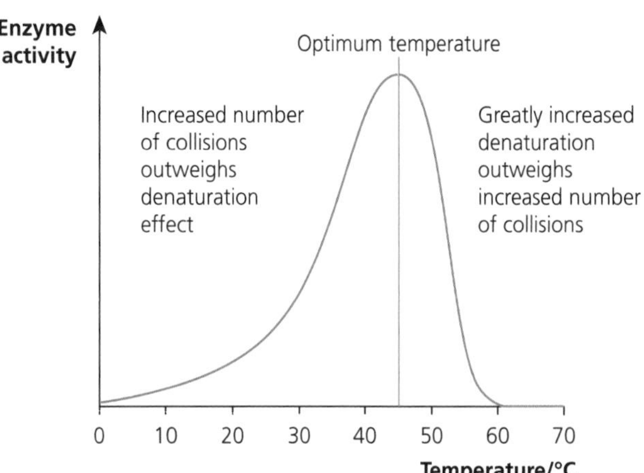

Figure 1.22 As temperature increases, the rate of reaction increases until the enzyme becomes denatured. It is often said that enzymes are denatured above 37°C, but most are more tolerant of heat. Individual enzymes have different optimums, but many are in the 45–55°C region

However, when the temperature gets too high, the enzyme molecules vibrate more vigorously and the weaker bonds, such as hydrogen bonds, break. The shape of the enzyme changes and the active site and substrate are no longer complementary. The enzyme is denatured and this process is permanent. Lowering the temperature will not make any difference — the enzyme will not revert back to normal.

Now test yourself

TESTED

12 Describe what happens when an enzyme is denatured. Make sure you use precise, A-level language.

Answer on p. 203

Nucleic acids are important information-carrying molecules

In this section we will look at the basic structure of the nucleic acids. The role of nucleic acids in making proteins is covered later in the specification and can be found on pp. 89–94.

The structure of DNA and RNA

REVISED

DNA

Deoxyribonucleic acid (DNA) has two vital properties:

- it holds the **genetic information** for making all the proteins that an organism needs
- it can copy itself exactly, time after time (replication — see p. 86)

You cannot have life without a molecule that can do this.

Deoxyribonucleic acid (DNA) is an information-carrying molecule that forms the genetic material in all living organisms.

DNA molecules are large **polynucleotide chains** formed when **nucleotides** bind together in a long chain. The bonds are formed by **condensation** and are called **phosphodiester bonds**. The polynucleotide chains in DNA are like a twisted ladder called a **double helix** (Figure 1.23). The molecule develops a natural twist that makes it more stable. The two sides of the ladder are made from alternating sugar–phosphate backbones, held together by **hydrogen bonds** between **base pairs** (Figure 1.24). There is one complete turn of the molecule for every 10 or so base pairs.

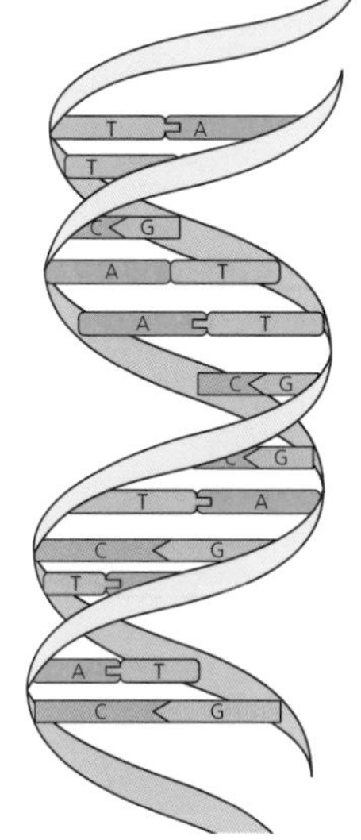

Figure 1.23 The double helix that forms when two polynucleotide strands twist around each other

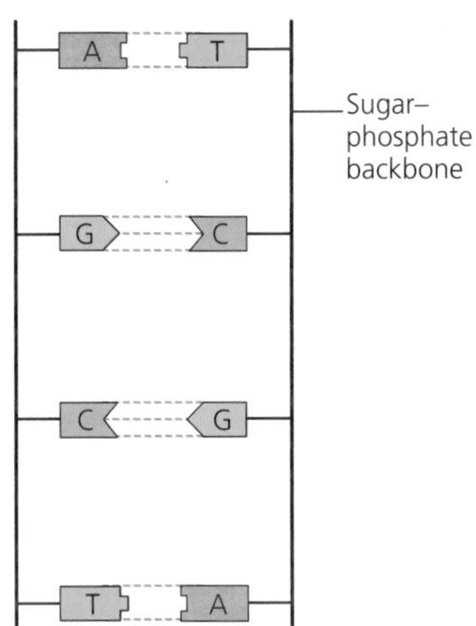

Figure 1.24 Part of a DNA molecule, with the sugar–phosphate backbones shown as single lines

Each nucleotide has three components (Figure 1.25):

- a **pentose** sugar (**deoxyribose**) — this is a five-carbon sugar, usually drawn as a pentagon
- a **phosphate group** — a PO_4^- ion that gives DNA a negative charge
- a **nitrogen-containing organic base** — one of four: **adenine** (A), **cytosine** (C), **guanine** (G) or **thymine** (T)

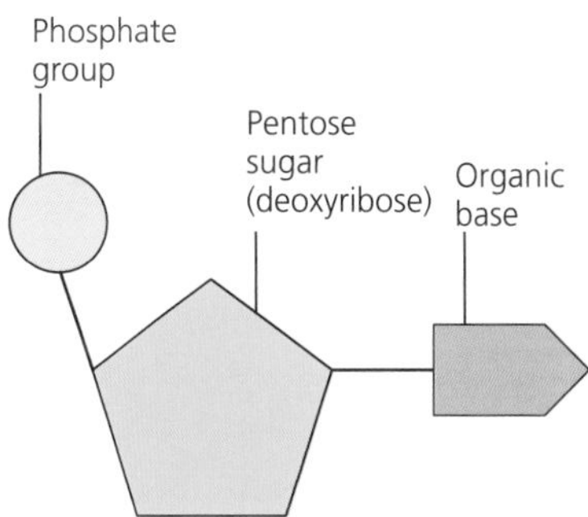

Figure 1.25 The structure of a single DNA nucleotide

Figure 1.26 shows a single polynucleotide strand. Note how the pentoses and the phosphates form a sugar–phosphate backbone.

Most bonds in DNA are covalent bonds, which are very strong and help to make DNA stable. However, the molecule cannot do its job if the bases do not come apart, so they are joined by relatively weak hydrogen bonds. The two strands of the DNA molecule must come apart for both replication and protein synthesis.

The two sides of the DNA molecule are joined by the bases, which are **specific**: A can only bond to T, and C can only bond to G. There are two hydrogen bonds between A and T (A=T) and three between C and G (C≡G).

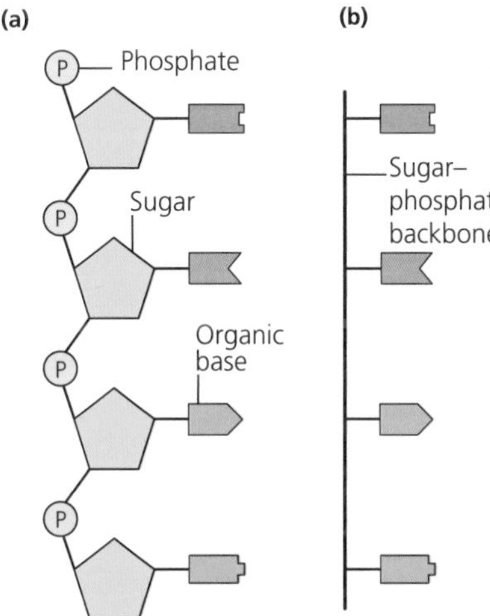

Figure 1.26 (a) Part of a single polynucleotide strand. (b) A simpler way to represent the same polynucleotide strand

Figure 1.27 shows part of a molecule of DNA. You should be able to identify the nucleotides, the sugar–phosphate bonds joining the nucleotides in one polynucleotide strand and the hydrogen bonds between **complementary** base pairs that are linking the two polynucleotide strands. As the two strands run in opposite directions, this is called antiparallel.

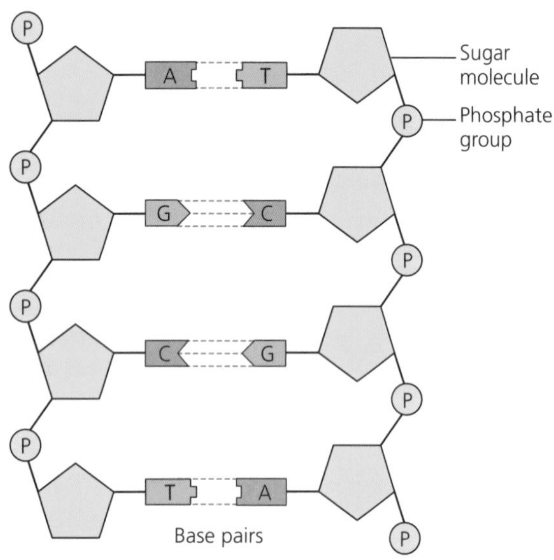

Figure 1.27 The overall structure of a DNA molecule

Exam tip

This section is just an introduction to DNA. You may see references to purines and pyrimidines, or 3′ and 5′ ends of DNA. These are not in the specification (and not in A2 either), so you are not expected to know about them for the exam.

Now test yourself

TESTED

13 If the base sequence on one strand of a DNA molecule reads AACTAGGTA, what will the base sequence read on the opposite strand?

Answer on p. 203

RNA

Ribonucleic acid (RNA) is a type of nucleic acid similar to DNA. There are three types of RNA.

There are three types of **ribonucleic acid (RNA)** molecule:

- messenger RNA (mRNA) — think of them as mobile copies of genes
- transfer RNA (tRNA) — to bring the right amino acids to the ribosome
- ribosomal RNA (rRNA) — makes up the body of the ribosome

You will learn about the roles of mRNA and tRNA later, on pp. 89–94.

Like DNA, RNA molecules are polymers of nucleotides, but there are key differences. RNA molecules:

- are single-stranded
- are shorter then DNA
- have the sugar ribose, not deoxyribose
- have the base uracil (U) instead of thymine (T)

DNA replication

REVISED

In the human body, most cells are not dividing so there is no need for the DNA to be replicated. However, if the cell is going to divide, the DNA must first be copied otherwise the resulting daughter cells will only contain half the genetic material. This ensures **genetic continuity** between generations of cells.

The key steps involved in DNA replication are as follows (Figure 1.28):

1 The enzyme **DNA helicase** attaches to the DNA molecule, breaks the **hydrogen bonds** between **complementary bases** in the two **polynucleotide strands** and **unwinds the double helix** of the DNA molecule, producing two single strands.

2 The enzyme **DNA polymerase** attaches to each strand. Each enzyme moves along its strand, catalysing the addition of new complementary **nucleotides** to the **exposed bases**. For example, wherever there is an exposed T base, an A base will be added. There is only ever one base that can be paired to the original. This is the key to keeping the code unchanged.
3 Sugar–phosphate bonds are formed and the double helix re-forms, so now there are two identical strands.

In each new molecule of DNA, half is original (it has been conserved) and half is new. This is what we mean by **semi-conservative replication**.

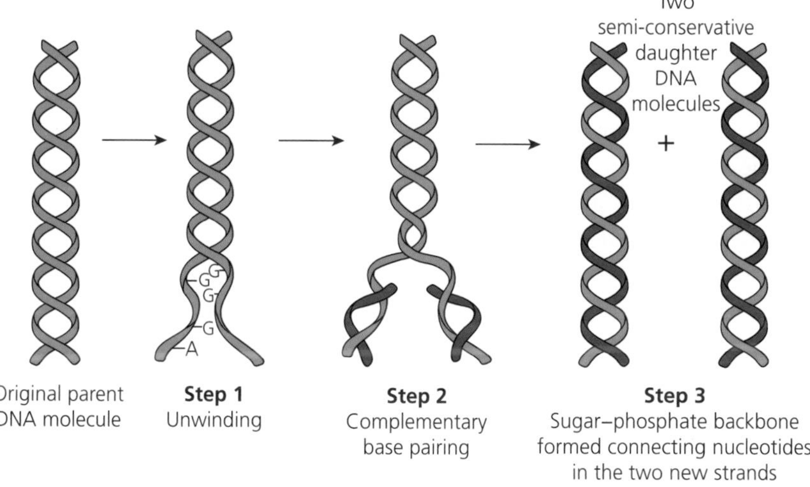

Figure 1.28 DNA replication

> **Typical mistake**
>
> Don't state that bases are added during DNA replication — it should be nucleotides. A nucleotide is a base with a sugar and a phosphate attached.

Now test yourself

TESTED

14 List four components needed for DNA replication.
15 Which bonds are broken by the enzyme DNA helicase?
16 In which part of the cell does DNA replication take place?
17 Explain what is meant by the term *semi-conservative replication*.

Answers on p. 203

ATP

The structure of ATP

REVISED

Adenosine triphosphate (ATP) is a substance found in all living organisms. Its function is to deliver instant energy in usable amounts. All organisms respire all the time because they need a constant supply of ATP.

> **Adenosine triphosphate (ATP)** is a substance, found in all living cells, that is involved in the transfer of energy.

Structurally, ATP is a **nucleotide**, like those that make up nucleic acids (Figure 1.29). It consists of:

- a pentose sugar (**ribose**)
- a base (**adenine**)
- three **phosphate groups**, which are the key to its function

ATP is a relatively simple molecule that releases its energy by **hydrolysing** into **adenosine diphosphate (ADP)** and an **inorganic phosphate group**, which can be written as PO_4^- or P_i.

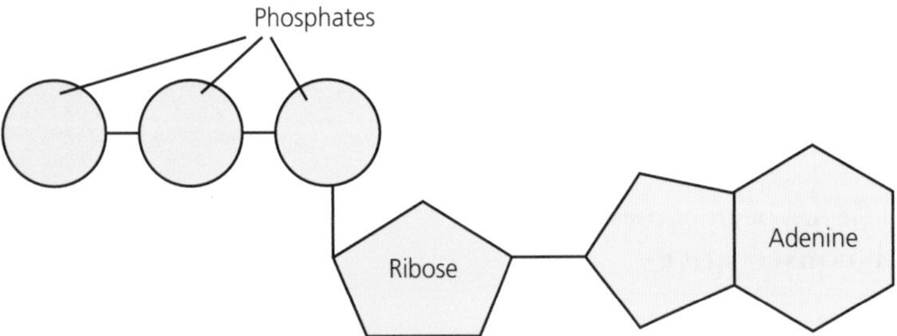

Figure 1.29 An ATP molecule

ATP:

- releases instant energy because it needs just one simple chemical step
- releases energy in usable amounts — if it released more energy than needed, the excess would be wasted as heat
- is a relatively small molecule so it can diffuse rapidly around the cell
- is often used to **phosphorylate** other molecules, making them more **reactive**
- is an unstable molecule that cannot be stored, so it must be resynthesised constantly

Typical mistake

Students often lose marks by writing P for phosphate. However, P is the symbol for the element phosphorus, not phosphate. You can use the word *phosphate* or the abbreviations P_i, PO_4^- or Ⓟ.

Exam tip

You don't need to memorise the detailed structure of ATP, just to understand how it provides energy.

ATP synthesis and uses

REVISED

ATP is made by the enzyme **ATP synthase**, a process that occurs in both **respiration** and **photosynthesis**. ATP is hydrolysed by **ATP hydrolase**, which is sometimes just called ATPase.

There is only a certain amount of ATP in a cell. It is constantly being broken down and needs to be resynthesised. In humans and all other warm-blooded animals that respire quickly, the weight of ATP produced each day is greater than the entire body weight. We have a relatively small amount of ATP but it is being constantly broken down and remade.

The balance of ATP and ADP/P_i in a cell is a bit like a battery. If it is all ATP, the battery is fully charged. If it is mostly ADP and P_i, the battery is run down and needs to be recharged by the process of respiration.

There are many processes that use ATP, but for exam purposes the three main ones are:

- muscular contraction
- active transport
- protein synthesis

Typical mistake

Don't say that ATP is needed *for* respiration. Respiration *makes* ATP.

Typical mistake

Don't say that ATP is a high-energy molecule or that it has high-energy bonds.

Water

Important properties

REVISED

Water is the most common component of **cells** and therefore of whole organisms. There is no life without water. It is so fundamental to life that the search for life on other planets centres around searching for those that might have liquid water.

Water molecules are simple and most substances of a similar size are gases (Figure 1.30). However, water is a liquid at most of the temperatures found on Earth because its molecules are polar — they have areas of positive and negative charge. As a consequence, they act like mini-magnets and cling to each other. Hydrogen bonds form between adjacent water molecules, giving water several important properties:

- It is a metabolite in many metabolic reactions, including condensation and hydrolysis.
- It is an important solvent. Substances with a charge will dissolve in water. Sugars, amino acids and ions such as sodium, chloride, calcium and potassium are all soluble in water.
- It has a relatively high heat capacity, meaning that it acts as a thermal buffer. It can absorb a lot of energy before the temperature rises and lose a lot before the temperature drops.
- It has a relatively large latent heat of vaporisation, meaning that when water evaporates it has a powerful cooling effect with little loss of water through evaporation.
- Water molecules are cohesive (they attract one another) and so they cling together, forming continuous columns that can withstand great tension. This is important in xylem vessels in plants.
- The cohesive nature also results in surface tension, where water meets air.

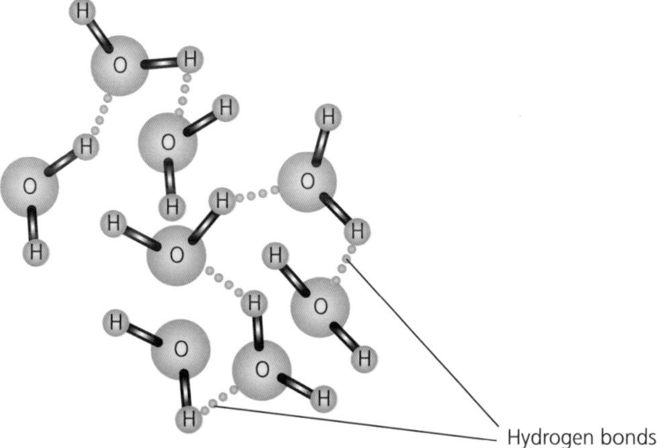

Figure 1.30 Water molecules

Inorganic ions

Important roles

REVISED

Inorganic ions occur in **solution** in the **cytoplasm** and **body fluids** of **organisms**. Some — such as sodium, potassium and chloride — occur in high concentrations. Others — such as iodine, copper and zinc — occur in very low concentrations.

Each type of ion has a specific role, depending on its properties. These are particularly important:

- **Hydrogen ions** (H^+) affect the **pH**. Acidic solutions have an excess of H^+ ions, whereas basic solutions have an excess of OH^- ions. pH is a logarithmic scale, so a change from pH 5 to pH 4 represents a ten-fold increase in the concentration of H^+ ions, whereas a change from pH 5 to pH 3 represents a hundred-fold increase. Enzymes are very pH sensitive (see p. 20).
- **Iron ions** (Fe^{2+}) are a vital component of **haemoglobin**, where they form the centre of the haem groups that combine with oxygen (see p. 66).
- **Sodium ions** are vital in the **co-transport** of **glucose** and **amino acids** (see p. 44). They are also vital in the transmission of nerve impulses.
- **Phosphate ions** are components of phospholipids, **DNA**, **RNA** and **ATP** (see pp. 21–25).
- The most common ions — sodium, potassium and chloride — are important in the regulation of water potential in cells and body fluids.

Exam practice

1 The two graphs show the effect of the two different types of inhibitor on the rate of an enzyme-controlled reaction.

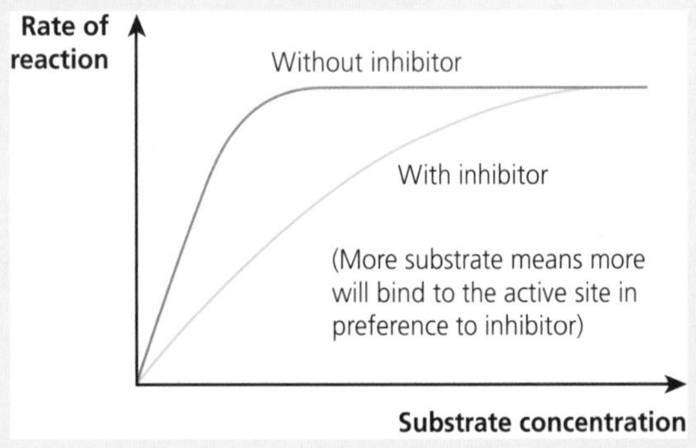

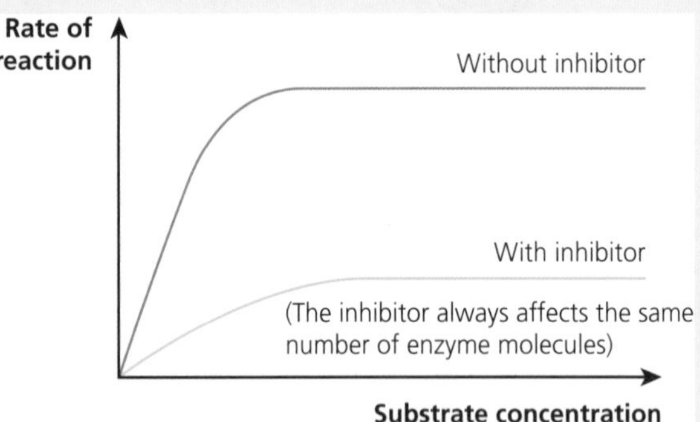

(a) Identify which graph shows the competitive inhibitor and which shows the non-competitive inhibitor. [1]

(b) Explain your answer to part (a). [2]

2 There are many different types of protein, but starch and glycogen are the same in all organisms. Explain why. [2]

3 Explain why you would not normally find glycogen in the blood of animals. [2]

4 A length of DNA was analysed and 21% of its bases were T (thymine). Calculate the percentage of the three other bases. [2]

Answers and quick quiz 1 online

ONLINE

Summary

By the end of this chapter you should be able to understand:

- That polymers are large molecules made from repeated monomers.
- The subunits that make up carbohydrates, lipids and proteins.
- How these subunits join to form larger molecules and the names of the bonds that join them.
- The importance of condensation and hydrolysis reactions.
- The way in which structure is related to function in starch, phospholipids and proteins.
- Why there is an infinite variety of proteins.
- The mode of action of enzymes.
- The factors that affect enzyme activity.
- The structure of DNA.
- The way in which DNA replicates.
- The differences between DNA and RNA.
- The structure of ATP and the way it releases energy.
- The properties of the water molecule and its importance in biology.
- The roles of inorganic ions, depending on their properties.
- The biochemical food tests for reducing sugars, non-reducing sugars, starch, lipids and proteins.

2 Cells

Cell structure

All living things are made from **cells**, which are tiny compartments of living tissue. Organisms such as bacteria, algae, yeast and amoeba are made from just one cell — they are **unicellular**. Large organisms are made from many cells — they are **multicellular**. Being multicellular is a huge advantage because it allows cells to specialise, which means that different cells have different functions.

Cell size

REVISED

In science, units of measurement usually go up or down in thousands, or 10^3:

- One thousandth of a metre is a millimetre (mm), or 10^{-3} m.
- One thousandth of a millimetre is a micrometer (μm), also called a micron, or 10^{-6} m.
- One thousandth of a micrometer is a nanometer (nm), or 10^{-9} m.

Cells and large organelles are measured in micrometers, whereas small organelles and molecules are measured in nanometers.

Exam tip

Remember:

- 1 mm = 1000 μm
- 1 μm = 1000 nm

Avoid using centimetres in science — it just causes confusion.

Example

Calculations in biology

The mitochondrion in the diagram is magnified 10 000 times. Calculate its actual size. [2]

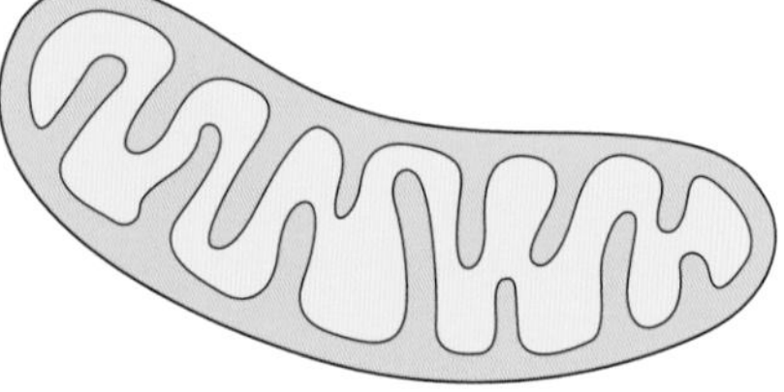

Answer

(a) Measure the diagram. It is 55 mm long. Don't use centimetres — you will get confused.
(b) Convert your measurement into micrometers. It is 55 000 μm.
(c) Divide your result by the magnification.

$$\frac{55\,000}{10\,000} = 5.5\ \mu\text{m}$$

Exam tip

You don't have to be a good mathematician to succeed in biology, but you do need to be numerate. You need to be able to do basic calculations and to recognise a silly answer when you see one.

Exam tip

Use this formula:

$$A = \frac{I}{M}$$

where:

A is the **A**ctual size of the object

I is the **I**mage size

M is the **M**agnification

You need to be able to rearrange this simple formula to work out *A*, *I* or *M*, depending on what the question requires.

Exam tip

Exam questions on scaling will sometimes give you a scale bar. You have to work out the magnification from the scale bar and then use it to work out the actual size of the object. For example, if the bar tells you that it represents 5 μm and you measure it as 10 mm on the page (|———|), you know that the magnification is 2000 (10 000/5). Think of it as 'How did 5 μm become 10 000 on the page? They must have enlarged it 2000 times.'

Two types of cell

REVISED

There are two basic types of cell: eukaryotic and prokaryotic.

Eukaryotic cells (Figure 2.1) are big and complex. Organisms made from eukaryotic cells include animals, plants and fungi. In fact, every organism apart from bacteria is made from one or more eukaryotic cells. You might think that all cells are tiny, but in the great scheme of things eukaryotic cells are big and complicated.

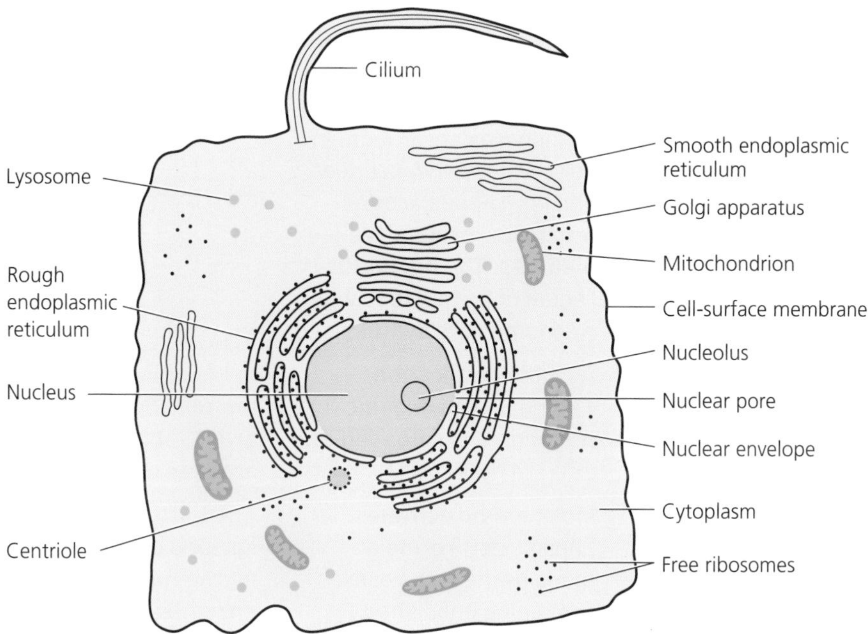

Figure 2.1 A eukaryotic cell (note that they do not all have cilia)

Prokaryotic cells (Figure 2.2) are small and simple. Bacteria have prokaryotic cells.

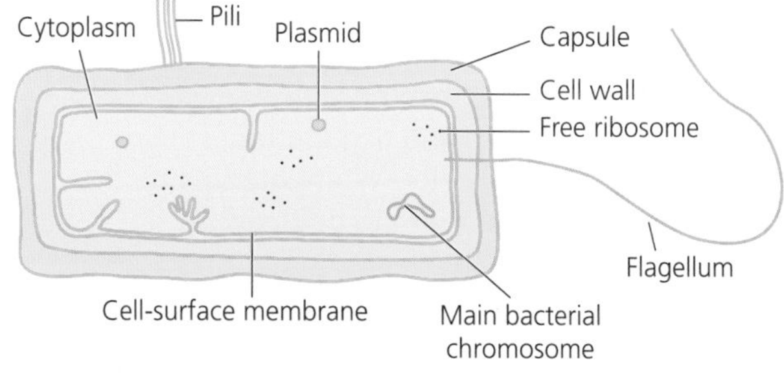

Figure 2.2 A prokaryotic cell

The structure of eukaryotic cells

REVISED

Animal cells

Table 2.1 outlines the features of an animal cell. In this example, we will look at an epithelial cell from the lining of the gut, which is adapted for the absorption of digested food molecules.

Table 2.1 The main organelles found in an animal cell

Organelle	What it looks like	What is does	Relating structure to function
Cell-surface (plasma) membrane		Controls what enters and leaves the cell. It is a chemical barrier that has no physical strength	A variety of proteins, embedded in the lipid bilayer, control the movement of specific substances
Nucleus		Holds the DNA. Unless the cell is dividing, there are no visible chromosomes — the DNA is spread out and is known as chromatin. In eukaryotes, DNA is linear and is wrapped around organising proteins called **histones**. The **nucleolus** is a region of the nucleus where ribosomes are synthesised	Many pores to allow substances in and out
Mitochondria		Aerobic respiration. The energy in organic molecules is released and used to make ATP	Folds of inner membrane (cristae) give a large surface area for making ATP
Golgi apparatus		Modifies and activates ('finishes off') proteins made by the cell. Vesicles containing unfinished proteins are constantly being added to the forming face, while vesicles containing the finished product are constantly being pinched off the maturing face	The flattened cavities contain all the necessary enzymes
Lysosomes		Small spheres of membrane containing digestive enzymes. Used for destroying organelles or whole cells. White blood cells use them to destroy engulfed bacteria	Bags of lytic enzymes
Ribosomes	Ribosome diameter = 10nm Large subunit mRNA	Protein synthesis. Proteins are assembled according to the codes on the genes	Holds together all the components of protein synthesis

Organelle	What it looks like	What is does	Relating structure to function
Rough endoplasmic reticulum (covered in ribosomes)	Ribosome	Transport system in the cytoplasm. Newly synthesised proteins are stored and packaged into vesicles	Extends throughout the cytoplasm. Large surface area for the attachment of many ribosomes
Smooth endoplasmic reticulum (no ribosomes)		Synthesis of lipids, including steroids. Breaks down various drugs and toxins, including alcohol	The cavities contain all the necessary enzymes

Plant cells

Plants cells are eukaryotes, so they have all of the features seen in animal cells such as a true nucleus, linear DNA, mitochondria and endoplasmic reticulum. However, in general, plant cells tend to be larger than animal cells and have three additional organelles:

- a **cell wall** made from cellulose — all plant cells have them
- a large, permanent **vacuole**
- **chloroplasts** in certain cells

Cell walls

Cell walls are secreted by the plant cell itself. They are soft at first, but then harden to prevent any further expansion. The way the cell wall is laid down can lead to the cell being a particular shape.

Healthy plant tissue is usually **turgid**, which means firm. Think of crisp lettuce leaves. Plant cells absorb water and the cytoplasm swells until it pushes out against the cell wall, which prevents any further expansion. This is a turgid cell — it is a bit like a properly inflated football where the bladder pushes out against the leather. If a plant is short of water, the cytoplasm shrinks and turgor is lost, resulting in the plant wilting.

Vacuoles

Vacuoles are storage organelles found in plant cells. They might store food, nutrients or water. They can also store waste products to protect the rest of the cell. By the time a plant cell has stopped growing, the vacuole is usually very large and it can hold large amounts of water, organic molecules or pigments.

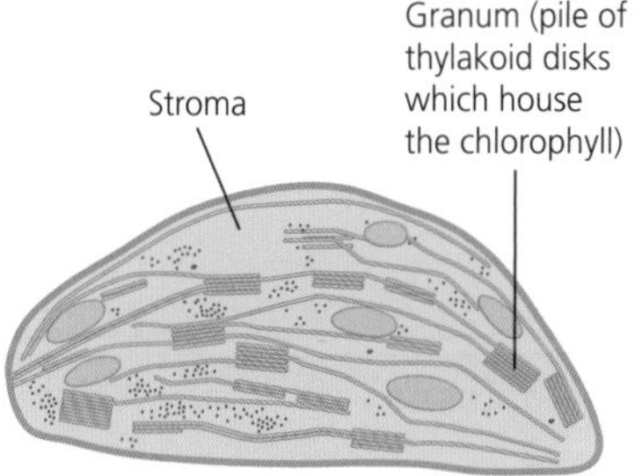

Figure 2.3 A chloroplast

Chloroplasts

Chloroplasts are organelles with one function — photosynthesis — the details of which are covered in the A2 course. Chloroplasts are found in some plant cells, usually those that are exposed to light (Figure 2.3).

Most of the plant's chloroplasts are concentrated in the palisade cells in the upper surface of leaves. The structure of a leaf allows the palisade cells to photosynthesise as efficiently as possible.

Chloroplasts are organelles that increase the surface area for the reactions of photosynthesis. Relating structure to function:

- chlorophyll molecules are attached to flat membranes called thylakoids
- stacks of thylakoids are called grana
- combining the two, there is a large surface area for light harvesting
- the flat, disk-like shape speeds up the diffusion of substances in and out

- the enzymes and other substances that turn carbon dioxide into sugar are found in the fluid (the stroma)
- starch grains allow sugars to be stored without lowering the water potential

Now test yourself

TESTED

1 Why are there no chloroplasts in the tissues of roots?

Answer on p. 203

Cells, tissues and organs

All plants and animals, by definition, are multicellular. The advantage of being multicellular is that cells can **specialise**, i.e. become adapted for different functions. Examples of specialised cells include neurones, spermatozoa and palisade cells.

A collection of specialised cells is known as a **tissue**. Examples of animal tissues include nerve, muscle and connective tissue. Plant tissues include palisade mesophyll, xylem and phloem.

An **organ** is a collection of tissues working together to perform a common function. Examples of animal organs include the eye, heart, lung and kidney, whereas plants have organs such as roots, leaves and flowers.

Organs are organised into **systems** that work together to perform a major function. Examples include the digestive, respiratory and reproductive systems.

Revision activity

Draw a table to summarise the similarities and differences between eukaryotic and prokaryotic cells.

The structure of prokaryotic cells (bacteria)

REVISED

If eukaryotic cells can be thought of as large and complex, prokaryotic cells can be thought of as small and relatively simple. The structure of prokaryotic cells is outlined in Table 2.2.

Table 2.2 The structure of prokaryotic cells

Feature	Details
Capsule	Some bacteria may have a layer of mucus outside the cell wall. This provides some protection from digestion by enzymes in the gut of animals and from drying out
Cell-surface (plasma) membrane	All living cells, including prokaryotic cells, have a cell-surface membrane
Cell wall	All bacteria have a complex cell wall, with more layers and components than the simple cellulose cell wall of plant cells. Prokaryotic cell walls contain a substance called murein (also known as peptidoglycan), which is a mesh formed from amino acids and sugars
Cytoplasm	A fluid in which most of the basic life functions occur, such as respiration
DNA	Bacterial DNA is always circular, i.e. the ends are joined together to form a loop. In every bacterial cell there is one main chromosome that contains all the genes essential for life, and several smaller plasmids that also contain useful genes. Bacterial DNA is not attached to organising proteins
Flagella (singular: flagellum)	A whip-like tail that is present in some prokaryotic cells to allow movement
Plasmid	Small loops of DNA. They contain genes that are useful rather than essential
Pili	Small, hair-like projections from the outer cell surface
Ribosomes	For protein synthesis. Bacteria have small, 70S ribosomes compared with the larger 80S ribosomes of eukaryotes

Viruses

REVISED

Viruses are tiny, infectious particles (Figure 2.4). They are usually classed as **non-living** because they are not made of cells — they are **acellular** — and they only do one of the seven signs of life: reproduction. They do not feed, respire, excrete or grow. They can only reproduce themselves by infecting a living cell and using the host's organelles to make more virus particles.

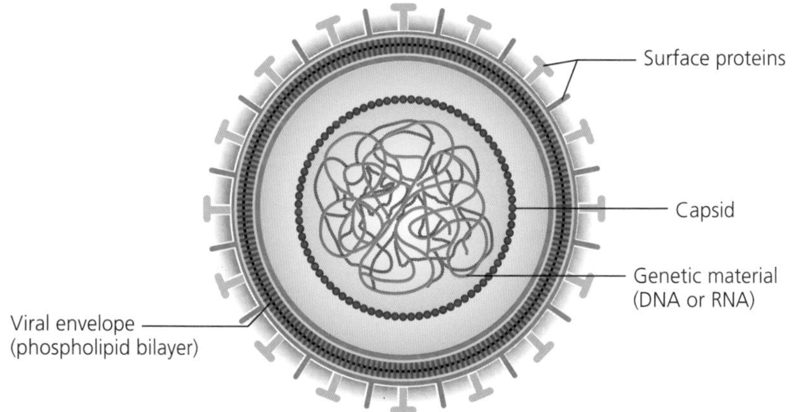

Figure 2.4 The basic structure of a virus

All viruses have:

- **genetic material** in the form of DNA or RNA. Viruses that contain RNA are called **retroviruses**
- an inner protein coat called a **capsid**
- **attachment proteins** that allow them to attach to end enter a cell

Now test yourself

TESTED

2 State whether each of these describes a eukaryotic or a prokaryotic cell.
(a) large; highly organised; lots of membrane-bound organelles
(b) has a complex cell wall
(c) contains endoplasmic reticulum
(d) has plasmids

Answer on p. 203

Methods of studying cells

REVISED

There are two basic types of microscope: optical (light) and electron.

Optical (light) microscopes

The microscopes used in schools and colleges are **light microscopes**, so called because they use light, focused by lenses. The power of microscopes is limited by the laws of physics, specifically the wavelength of light and electron beams. Most light microscopes have a maximum magnification of a few hundred and even the most expensive will not go much beyond 1500×.

Electron microscopes

Electron microscopes have greater magnification and resolution than light microscopes. They use a beam of electrons, focused by magnets. The wavelength of an electron beam is much smaller than the wavelength of light, so its potential to enlarge and show detail is much greater.

Exam tip

In questions about resolution, there is often a second mark to be gained by including a bit of physics: the wavelengths of electron beams are smaller than the wavelength of light.

There are two types of electron microscopes commonly used in biology: **transmission electron microscopes** (TEMs) and **scanning electron microscopes** (SEMs). TEMs were developed first and work on the principal that electrons are transmitted through the specimen. With SEMs, electrons bounce off the surface of the specimen and images are produced by computers.

TEMs produce images that are two-dimensional. Specimens need to be thin, coated with a heavy metal (an electron-dense material) and observed in a vacuum. This means that it takes a while to prepare specimens and you cannot observe living material. On the other hand, SEMs are can study solid specimens, not just thin sections. They use computers to produce three-dimensional images. The resolution of TEMs is slightly higher than that of SEMs.

The best electron microscopes can magnify objects by as much as 10 000 000× and have a resolution of less than 1 nm. This means that at maximum power they can distinguish between individual molecules.

Now test yourself

TESTED

3 Draw a table of the differences between transmission electron and scanning electron microscopes.

Answer on p. 203

Magnification and resolution

Many candidates confuse magnification and resolution.

- **Magnification** is how many times the image has been enlarged. For example, if an image says 20 000×, it is 20 000 times larger than it is in reality. Magnification can be calculated using this formula:

$$\text{magnification} = \frac{\text{size of image}}{\text{size of real object}}$$

See p. 29 for more information on how to use this formula.

- **Resolution** means detail. It is defined by how close together two objects have to be before they are seen as one. For example, a resolution of 1 μm means that two objects separated by less than 1 μm will appear to be one object.

Magnification is the ratio of the image size to the object size.

Resolution is the ability to distinguish two separate points that are close together.

Typical mistake

When talking about electron microscopes, candidates often mention magnification but forget about resolution, which is the key property.

Cell fractionation and ultracentrifugation

Cell fractionation involves separating the different components of cells. For example, you might want to study the mitochondria in a piece of liver or the chloroplasts in spinach. How do you get samples of pure organelles? The basic process is as follows:

1 Add an ice-cold, isotonic buffer to the tissue. When cells are broken up, chemicals that do not normally come into contact will mix, so a solution is needed that is:
   - **ice cold** to minimise unusual reactions, especially ones involving enzymes
   - **isotonic** to prevent organelles swelling up or shrinking due to osmosis
   - a **buffer** to prevent any damaging pH changes

2 Homogenise the piece of tissue (put it into a blender). This breaks up the cell membranes and cell walls, if present, so you get a 'soup' of organelles.

3 Filter the resulting mixture to get rid of any large lumps of unbroken cells and cell debris.

4 Spin the mixture in a centrifuge. This creates a super-gravitational field that causes organelles to separate according to their density. The

mixture is spun relatively slowly at first. This causes the nuclei to sediment out, forming a solid **pellet** of sediment at the bottom of the tube. The rest of the organelles are still suspended in the fluid, called the **supernatant**.

5 If the supernatant is spun again at higher speed (**ultracentrifugation**), the next densest organelles sediment out, which are the mitochondria. Increasing the speed separates out the lysosomes, endoplasmic reticulum and finally the tiny ribosomes.

All cells arise from other cells

Mitosis

REVISED

For multicellular organisms to grow and develop, cells must **divide** and then specialise. **Mitosis** is cell division that involves the **parent cell** dividing to produce two identical **daughter cells**, each with the **identical copies** of **DNA** produced by the parent cell during **DNA replication**. This means the cells are genetically identical to the parent cell and to each other.

Mitosis is the division of cells to produce two genetically identical daughter cells.

Mitosis is divided into four key stages, as shown in Table 2.3.

Table 2.3 The four stages of mitosis

Phase	What it looks like	Main events
Prophase	Centromere Spindle fibres Two chromatids make up each chromosome	Chromosomes coil and become shorter and fatter, so that they are now visible with an optical microscope The nuclear envelope disappears Protein fibres form a **spindle** in the cell The spindle fibres attach to the middle (**centromere**) of each chromosome
Metaphase		The spindle fibres pull the chromosomes to the middle (equator) of the cell
Anaphase		The centromere holding each pair of sister **chromatids** together divides The spindle fibres shorten and pull the chromatids to opposite poles of the cell The chromatids are now called chromosomes again
Telophase		The two sets of chromosomes group together at each pole and a nuclear envelope forms around each group The chromosomes uncoil, becoming chromatin again They are no longer visible with an optical microscope Finally, the cytoplasm divides, resulting in two new cells. This process is called **cytokinesis**

Typical mistake

In questions about mitosis, candidates often make vague references to chromosomes. Try to use specific language such as *chromatid, spindle, centromere* and *pole*.

Exam tip

Make sure that you can recognise each stage of mitosis and describe what is happening.

The cell cycle

REVISED

The complete life of a cell, from the moment it is created until it splits again into two new cells, is called the **cell cycle**. Mitosis is only a small part of the cell cycle. The rest is known as **interphase**, when the chromosomes are copied and the genetic information is checked. During interphase, the cell also increases in size, produces new organelles and stores energy for another division. If the cell is going to divide again, DNA is replicated during interphase, just before the chromosomes condense and become visible.

Most cells are in interphase for most of the time. During interphase:

- the nucleus is intact — its membrane can be seen
- chromosomes are not visible — the DNA is spread out
- genes are being expressed — used to make proteins

The cell cycle is divided into four phases (Figure 2.5):

1. G_1 phase — the cell increases in volume as new cytoplasm is made. The cell might prepare to replicate its DNA.
2. S phase — if the cell is going to divide, it enters the S phase in which it synthesises (replicates) its DNA.
3. G_2 phase — the cell continues to grow, and synthesises the enzymes and structures needed for mitosis.
4. M phase — the nucleus of the cell divides by mitosis.

The G_1, S and G_2 phases are collectively known as interphase. Once mitosis is complete, the cell divides into two cells during **cytokinesis** (Figure 2.6).

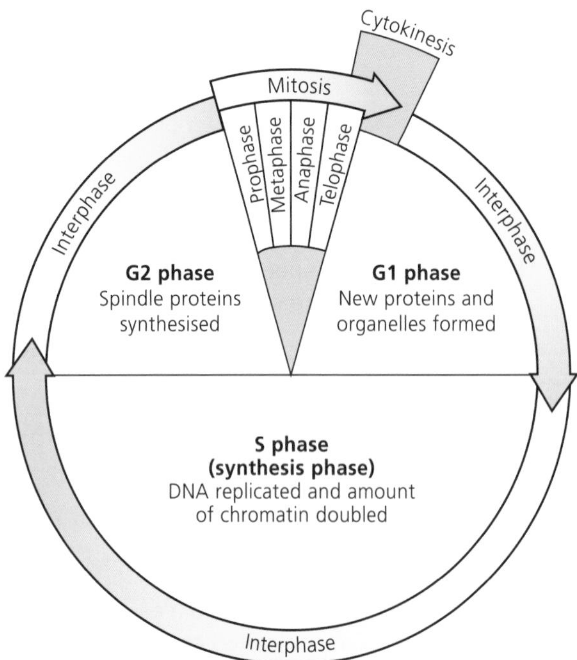

Figure 2.5 The cell cycle

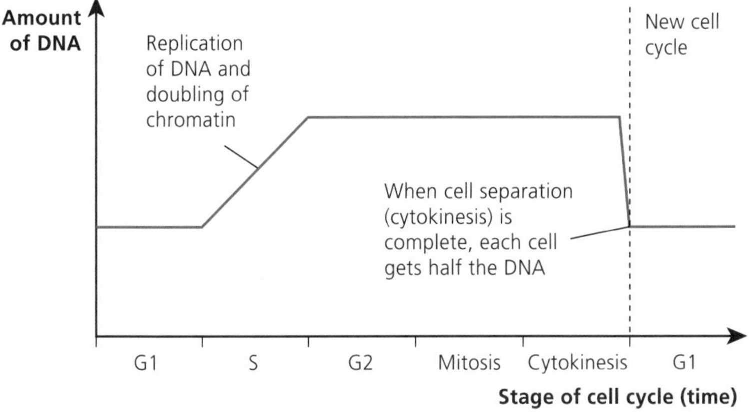

Figure 2.6 Changes in the amount of DNA during the cell cycle

Now test yourself

TESTED

4 How many different nucleotides must be available in the cytoplasm in order for DNA replication to proceed?

Answer on p. 203

Cancer: when mitosis goes wrong

REVISED

Mitosis is a carefully controlled process so that cells only divide when they should. When the genes that control the cell cycle mutate, the result is uncontrolled division, resulting in a swelling or growth known as a **tumour**. There are two main types of tumour:

- **Benign** tumours are usually slow-growing and do not spread to other parts of the body. These are not classed as cancer.
- **Malignant** tumours grow much more quickly and often spread. Cells can break off and start up new tumours in other parts of the body. It is these tumours that are classed as **cancer**.

Cancerous cells go through the cell cycle faster than normal cells, so disrupting the cell cycle is an obvious target. The problem is that anything that affects cancerous cells also affects normal cells.

Broadly, there are three approaches to **cancer treatment**:

- Surgery — removal of the tumour. This may be difficult because of where it is located (e.g. in the head) and because it is difficult to remove all of the cancerous cells.
- Radiotherapy — radiation damages DNA.
- Chemotherapy — drugs are used to kill cancerous cells, preventing them from dividing or damaging them so that they kill themselves. There are several approaches to chemotherapy. Examples of drug action include:
  - blocking the enzymes involved in DNA synthesis in the G_1 phase
  - preventing DNA from unwinding so that replication is impossible
  - inhibiting the synthesis of new nucleotides
  - preventing the development of the spindle

The mitotic index

The mitotic index (*MI*) is a ratio showing the proportion of cells undergoing mitosis in a piece of tissue.

$$MI = \frac{\text{number of cells undergoing mitosis}}{\text{number of cells in sample}}$$

In practice, the number of cells undergoing mitosis is taken to be the number of cells that have visible chromosomes. This index is useful for determining when a tissue is becoming cancerous and for assessing the effectiveness of cancer treatment.

Exam tip

It is easy to get carried away and study cancer in too much detail. The specification states that 'Many cancer treatments are directed at controlling the rate of cell division'. If you have learnt the cell cycle and can make sense of graphs, you will be fine with questions about cancer treatments.

Binary fission in prokaryotic cells

REVISED

Bacterial reproduction is amazingly rapid and, in the right conditions, it can happen as often as every 10 minutes. It basically involves cells splitting in half and is known as **binary fission**. It involves:

- replication of the circular DNA and of plasmids
- division of the cytoplasm to produce two daughter cells, each with a single copy of the circular DNA and a variable number of copies of plasmids

Viral reproduction

REVISED

Being non-living and non-cellular, **viruses** do not undergo cell division. They can reproduce only by entering a host cell and taking over some of the cell's organelles, using them to make more virus particles. All viruses are basically composed of a protein capsule that surrounds some nucleic acid, either DNA or RNA. The genes contained in the nucleic acids provide the information for making new virus particles. Once complete, the new virus particles burst out of the cell and can re-infect new cells.

Transport across cell membranes

The fluid-mosaic model of membrane structure

REVISED

The basic structure of all cell membranes, including cell-surface membranes and the membranes around the organelles of eukaryotes, is the same. The **fluid-mosaic model** (Figure 2.7) is used to describe this structure. Described as 'protein icebergs in a lipid sea', the key elements are:

- a bilayer of **phospholipid** molecules
- cholesterol, which reduces the permeability and fluidity of the membrane, making it more stable
- **proteins** that float in the phospholipid bilayer. Some proteins are partially embedded in the bilayer — these are called extrinsic proteins. Others span the bilayer — these are called intrinsic proteins. Some proteins float freely in the bilayer, whereas others may be bound to other components in the membrane or to structures inside the cell
- **glycolipids** (sugars attached to lipids) and **glycoproteins** (sugars attached to proteins), which function in cell signalling or cell attachment

The **fluid-mosaic model** is the basic structure of all membranes in a cell. Whenever you draw a cell, each line is a membrane.

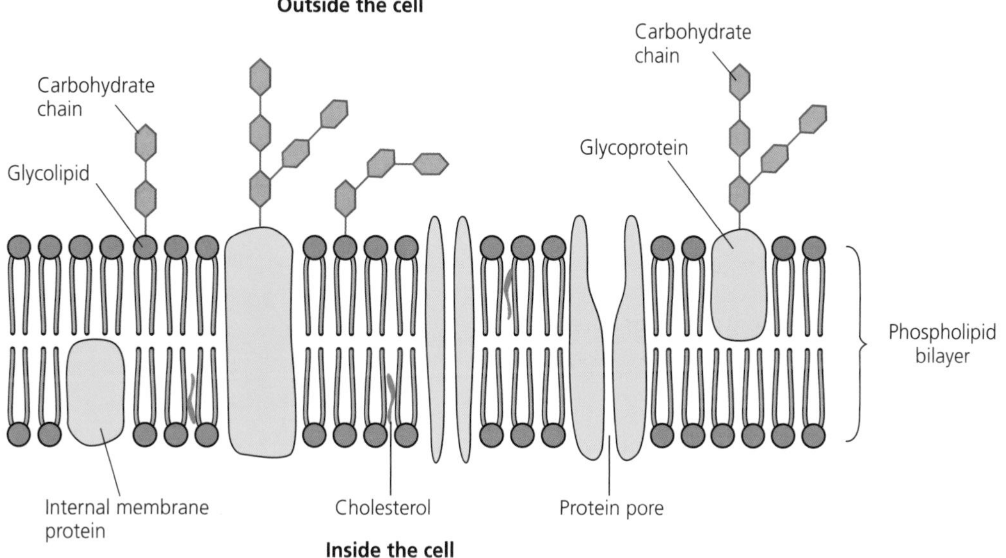

Figure 2.7 The fluid-mosaic model of membrane structure

What can and cannot get across membranes?

REVISED

Small, simple molecules such as water, oxygen and carbon dioxide can pass freely across. Lipid-soluble molecules such as alcohol (ethanol), aspirin, steroid hormones and some vitamins can pass through the phospholipids. Small, water-soluble molecules such as glucose and amino acids can pass through protein pores. Ions such as sodium, potassium and chloride pass through specific membrane proteins. Large molecules such as proteins cannot pass through at all, except by endocytosis and exocytosis.

The difference between lipid-soluble molecules and water-soluble molecules is important in biology. As a general rule, ions and polar molecules (those with areas of positive and negative charge) will dissolve in water. Molecules that do not have a charge will dissolve in lipid.

Now test yourself

TESTED

5 Suggest how fructose passes across cell membranes.

Answer on p. 203

Diffusion

REVISED

Diffusion is the **passive transport** of substances down a **concentration gradient**, until they are evenly distributed.

> **Diffusion** is the net movement of molecules down a concentration gradient.
>
> A **concentration gradient** exists between an area of high concentration and an area of low concentration.

In liquids and gases, particles are free to move and are constantly colliding with each other and changing direction (Figure 2.8). As a consequence, particles bounce around until they are evenly distributed. Diffusion is vital in gas exchange.

Examples of diffusion include:

- in the lungs, oxygen diffuses from the air into blood and carbon dioxide diffuses in the opposite direction
- in the gills, oxygen diffuses from water into the blood
- in leaves, carbon dioxide diffuses into palisade cells

The rate of diffusion is affected by:

- surface area
- differences in concentration (or gradient) between the two areas
- thickness of the exchange surface between the two areas — often, this is the thickness of the membrane
- temperature — at higher temperatures, particles bounce around and spread more quickly

Fick's law is used to measure of the rate of diffusion. It states that:

$$\text{rate of diffusion} \propto \frac{\text{surface area} \times \text{concentration gradient}}{\text{thickness of the exchange surface between the two areas}}$$

In order for diffusion to be as fast as possible, the two top factors need to be as large as possible and the one at the bottom needs to be as small as possible. Looking at the lungs, for example:

- lots of alveoli provide a large surface area
- constant breathing and a good blood supply maintain the concentration gradient
- thin alveolar walls minimise the thickness of the exchange surface, resulting in a short diffusing pathway

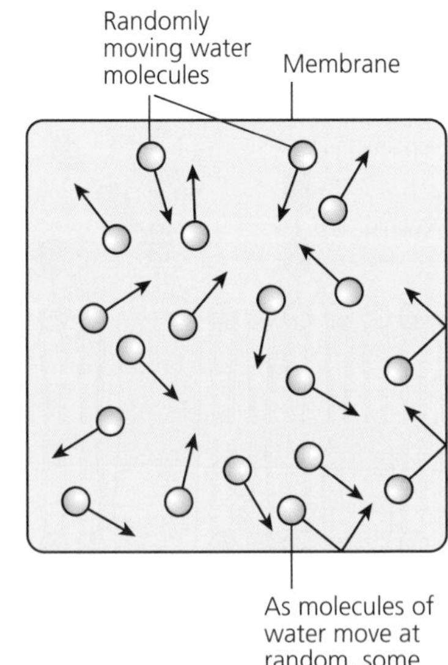

Figure 2.8 Like all particles that are free to move, water molecules bounce around at random, colliding and changing direction until they are evenly distributed

Now test yourself

TESTED

6 Explain why diffusion is faster at high temperatures.

Answer on p. 203

Facilitated diffusion

REVISED

Facilitated means 'helped' or 'speeded up'. **Facilitated diffusion** is diffusion helped by a specific protein in the membrane. There are two basic types:

1. **Carrier proteins** transport medium-sized molecules into and out of cells. To do this, they usually have to undergo a change in shape. For example, the entry of glucose into the cells of the body is speeded up by having specific glucose carrier proteins.
2. **Channel proteins** have a hole running through the middle. They transport ions into and out of cells. Some are specific and only allow certain ions such as sodium or chloride to pass through; others are non-specific and allow several ions to diffuse. Some are open all the time and others have gates (gated channels) that can open and close.

Facilitated diffusion stops when equilibrium is reached. It *does not* move substances against a concentration gradient; nor does it need energy in the form of ATP.

Osmosis

REVISED

Osmosis is a special case of diffusion in which water moves from a solution of higher water potential to a solution of lower water potential through a partially permeable membrane.

All substances that dissolve in water do so because they attract water molecules. A glucose molecule, for example, is surrounded by a layer of water molecules. The more glucose molecules there are in a solution, the more water molecules they will attract.

Osmosis is the net movement of water from a region of higher water potential to a region of lower water potential through a partially permeable membrane.

Osmosis happens whenever two solutions are separated by a membrane that allows the water molecules (but not the solute) to move. If there were no membrane, the solute would simply diffuse until it was evenly distributed. However, the presence of a partially permeable membrane makes all the difference — if the solute cannot move into the water, the water will move into the solute.

Exam tip

Osmosis is the *net* movement of water. Water molecules will move in both directions but, overall, there will be more moving in a particular direction.

Water potential

Water potential is a measure of tendency of water molecules to move from one place to another. Imagine a beaker of pure distilled water. It has the highest possible water potential — zero. Add a spoonful of sugar and you will lower the water potential, i.e. make it more negative. It is a negative scale because all solutions have a lower water potential than pure water. The more sugar you add, the lower the water potential gets. Really concentrated syrupy solutions have very low water potentials (Figure 2.9).

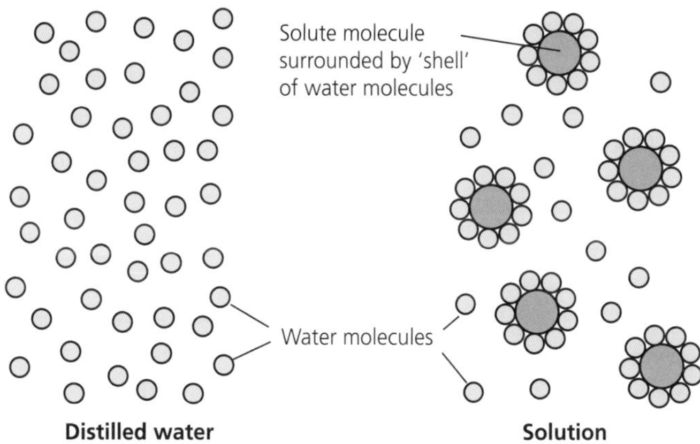

Figure 2.9 The more solute there is, the lower the water potential

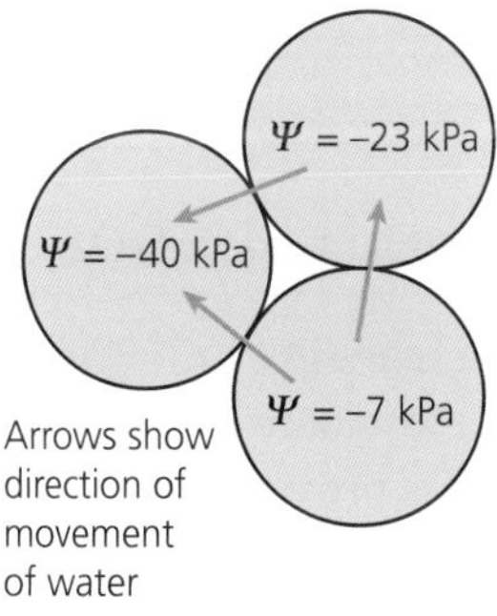

Figure 2.10 When cells of different water potentials are in close contact, water will pass to the ones with the lower values

Water potential is given the Greek letter (*psi*, pronounced 'sigh'). It is a measure of pressure and is measured in kilopascals (kPa). A weak solution might have a water potential of −50 kPa, whereas a more concentrated one might have a water potential of −300 kPa (Figure 2.10).

Now test yourself

TESTED

7 A cell with a water potential of −200 kPa is next to one with a water potential of −250 kPa. Which way will the water molecules move? Explain your answer.

Answer on p. 203

Osmosis in cells

Osmosis is vital in biology because all cells contain a solution (cytoplasm) and they are all surrounded by a solution. In the case of mammals, all cells are bathed in tissue fluid. Cell membranes are freely permeable to water but not to solutes.

Animal cells

Think about what would happen if we added pure water to a sample of blood. The cytoplasm in the red cells has a lower water potential than the surrounding solution, so the cells would absorb water, swell up and burst. Animal cells have no cell wall and so burst easily.

Exam tip

You need to be able to answer osmosis questions in terms of water potential. Don't talk about 'more concentrated solutions' or 'less concentrated solutions'.

If we add sea water to a sample of blood, the opposite happens. Sea water is a concentrated solution of salt and it has a lower water potential than the red blood cells. Water is drawn out of the cells so that they shrivel up.

All of the cells in our body behave in the same was as red blood cells. Animal cells need to be bathed in a solution that is isotonic, which means that it has the same water potential as the cytoplasm. That is why we have kidneys — to get rid of the excess water or to conserve water when we have too little.

Plant cells

Plant cells are different from animal cells because they have a cell wall, so they cannot burst. If plant cells are placed in pure water, they absorb water and swell until the cell wall prevents any further increase. In this state the plant cell is said to be turgid. This is the normal, healthy state for most plant cells. Plant cells are covered in more detail on p. 32.

Now test yourself TESTED

8 Most animals have organs such as kidneys to get rid of excess water. Explain why plants do not need these.

Answer on p. 203

Active transport

REVISED

Active transport moves molecules against a concentration gradient — from a low concentration to a higher one. This can only happen if energy is provided to drive the process. This energy is released from the **hydrolysis of ATP** (produced in respiration). Active transport also requires specific membrane proteins, usually referred to as **carrier proteins** or **pumps**.

Active transport is the movement of molecules using metabolic energy in the form of ATP.

Some cells are adapted to maximise the rate of active transport. One example is the epithelial cells that line the small intestine, which show these adaptations:

- Microvilli are folds/projections in the cell-surface membrane. The more membrane there is, the more carrier proteins and channel proteins there are for active transport.
- Many mitochondria provide the ATP for active transport.

Exam tip

Candidates often lose marks by saying that cells have villi, but the correct term is *microvilli*.

Now test yourself TESTED

9 If there were 100 molecules of a vitamin in the gut, how many molecules could be absorbed into the blood by the following means?
(a) facilitated diffusion
(b) active transport
10 (a) Which process in the cell produces ATP?
(b) In which organelles does this process mainly take place?

Answers on p. 203

Typical mistake

Candidates often state that epithelial cells have 'thin membranes', but they have normal membranes. It is the cells themselves that are thin, creating a short diffusion pathway.

Exam tip

If a question about cell transport says that a process is inhibited by a respiratory poison such as cyanide, you know it is an active process that needs ATP. Cyanide does not affect diffusion, facilitated diffusion or osmosis, but it does affect active transport.

Co-transport and the absorption of glucose

REVISED

By far the most common monosaccharide in our diet is **glucose**, which is absorbed by cells lining the mammalian ileum (the small intestine) into the blood by a process known as **co-transport**. It involves both facilitated diffusion and active transport, and absorbs **sodium ions** (Na^+) at the same time (Figure 2.11).

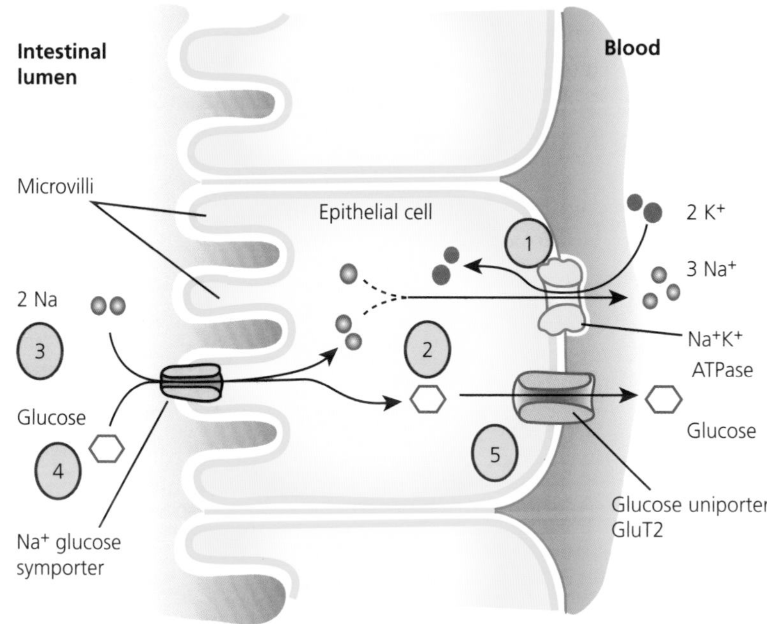

Figure 2.11 The absorption of glucose

1. Sodium is pumped out of the cell and into the blood by active transport.
2. This creates a low concentration of sodium inside the cell.
3. Sodium binds to the Na^+ glucose symporter (it is facilitated diffusion of sodium).
4. When sodium binds, so does glucose, so glucose is transported into the cell along with sodium.
5. There is a higher concentration of glucose inside the cell than outside, so glucose passes into the blood by facilitated diffusion (through the GluT2 membrane protein).

Now test yourself

TESTED

11 Explain what is meant by the term *co-transport*.

Answer on p. 203

Cell recognition and the immune system

Cell recognition

REVISED

All cells have proteins and glycoproteins on their surface. These molecules allow the cells of the immune system to tell the difference between self and non-self. Specifically, they allow the immune system to recognise:

- **pathogens** — disease-causing organisms such as bacteria and viruses
- **cells from other organisms** of the same species. Every individual has a unique combination of proteins on their cells, so a transplant from another individual will be recognised as foreign
- **abnormal body cells**. The immune system can recognise damaged cells, such as cancerous cells, and destroy them before a tumour develops
- **toxins** — potentially dangerous by-products of the metabolism of pathogens

Defence against disease

REVISED

Defence against disease takes two forms:

- preventing the entry of pathogens into the body
- the **immune response**, which combats pathogens that have already entered the body

The immune response can be divided into two categories:

- **Non-specific responses** take place whatever the type of pathogen that gets in. The main non-specific responses are fever, inflammation (when an area becomes red, painful, hot and swollen) and phagocytosis.
- **Specific responses** involve the immune system recognising the type of pathogen that has entered and making exactly the right type of antibody to deal with it.

Now test yourself

TESTED

12 List four of the body's barriers that prevent the entry of pathogens.

Answer on p. 203

Antigens and antibodies

Antigens are substances not normally found in the host's body, which stimulate the immune system into action. More specifically, antigens stimulate the production of a corresponding **antibody**. Antigens are usually proteins, polysaccharides or combinations of the two (glycoproteins). Non-self cells are recognised by the body because they have antigens on their cell-surface membranes. This means that phagocytes know which cells to attack. Pathogens are recognised because they are covered in antigens.

Antigens are molecules on the surface of cells that trigger an immune response.

An **antibody** is a protein made by lymphocytes in response to particular antigens. They have specific binding sites and are capable of acting against the pathogen.

Now test yourself

TESTED

13 Would glucose be a likely molecule to act as an antigen? Explain your answer.

Answer on p. 203

All antibodies are Y-shaped proteins (Figure 2.12). Their molecules have:

- four polypeptide chains, two long and two short
- a variable region on the ends of the chains
- two antigen-binding sites, formed by the variable regions, that are complementary to the antigen

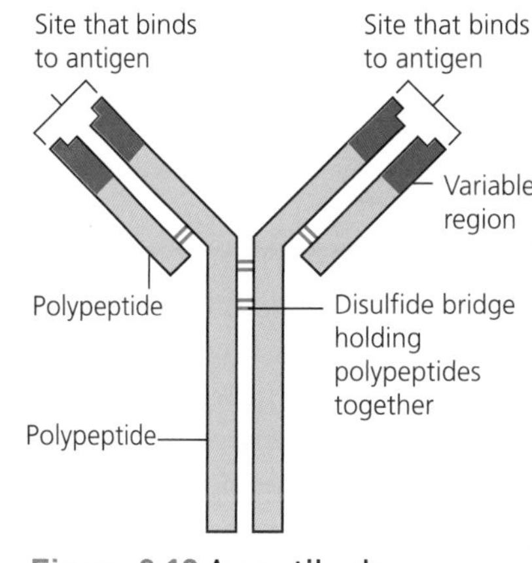

Figure 2.12 An antibody molecule

Exam tip

Remember two key points:

- antibodies consist of four polypeptide chains, so they have a quaternary structure
- different amino acid sequences give a different shape to the variable region

Antigen–antibody complexes form, which can neutralise the pathogen as follows:

- they label the pathogen as foreign — leading to phagocytosis
- they coat the pathogen so that it cannot invade a host cell
- they make pathogens stick together (**agglutination**) — again, preventing invasion of host cells

Now test yourself

TESTED

14 Explain the differences between antigens and antibodies.
15 Antibodies can be made only by white blood cells. Why is it not possible to simply make them in the lab, like aspirin or paracetamol?

Answer on p. 204

Phagocytes and lymphocytes

REVISED

The central component of the immune system is the vast array of white cells, which also have the general name *leucocytes*. The immune system is complex and it is easy to learn too much detail. You only need to know about these cells:

- **Phagocytes** (Figure 2.13). These cells usually have a lobed nucleus and carry out phagocytosis.
- **Lymphocytes** (Figure 2.14). These cells make antibodies. There are two basic types, **B lymphocytes** (or B cells) and **T lymphocytes** (or T cells). All white cells originate in bone marrow, but B cells are so called because they mature in bone marrow, whereas T cells mature in the thymus gland (in the chest). There are lots of different types of T cell, but the only two you need to know about are **helper T cells** (or TH cells) and **cytotoxic T cells** (also known as TC cells or killer T cells).

Exam tip

There are lots of other types of white cells including macrophages, neutrophils and basophils, but you don't need to know about these.

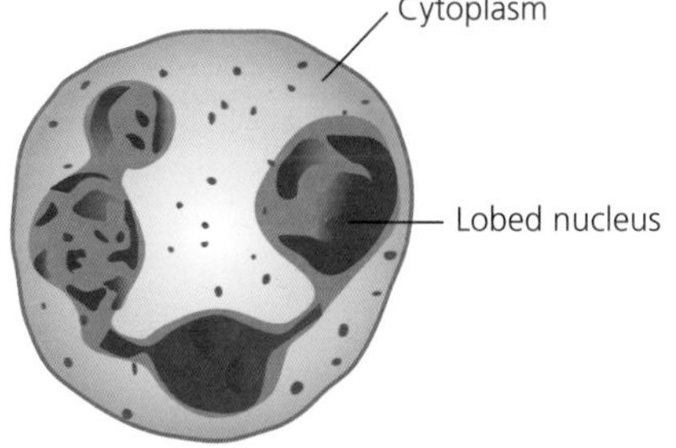

Figure 2.13 Phagocytes can be recognised by their lobed nucleus

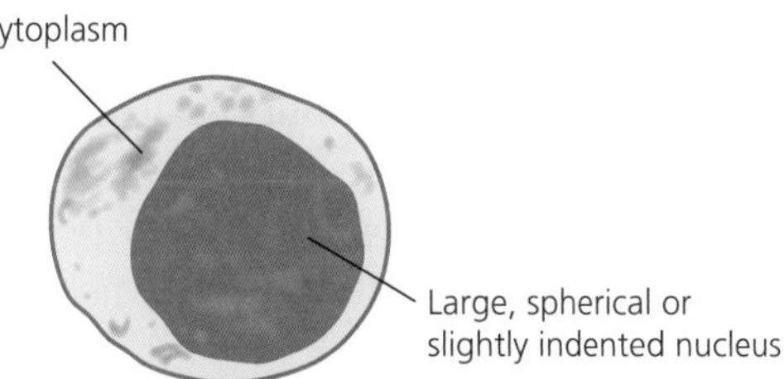

Figure 2.14 Lymphocytes can be recognised by their large round or kidney-shaped nucleus. T and B cells look exactly the same

Phagocytosis and the role of lysosomes

REVISED

Phagocytosis is a process in which white cells called phagocytes engulf and destroy bacteria and other foreign material that gets into the body (Figure 2.15). The steps are as follows:

1 The phagocyte recognises and engulfs the bacterium. The cytoplasm flows around and joins on the other side, leaving the bacterium enclosed in a vacuole called a phagosome.
2 **Lysosomes** move to the phagosome. Lysosomes are small vesicles (bags) of digestive enzymes.
3 The lysosome membrane fuses to the phagosome membrane to form a **phago-lysosome**. The enzymes digest the bacterium.
4 The remains of the bacterium are ejected from the cell, via exocytosis.

At the site of infection, phagocytes can engulf many bacteria. Eventually, a mixture of bacteria, dead tissue and dead white cells form a creamy fluid called pus.

Phagocytosis is when a phagocyte engulfs and ingests a pathogen.

Exam tip

Make sure you have watched an animation showing phagocytosis.

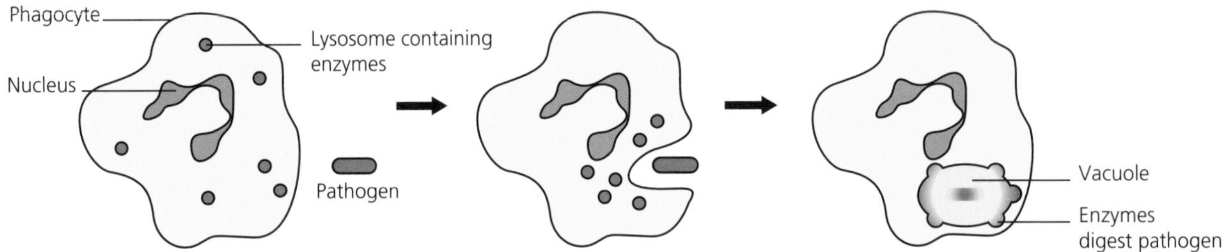

Figure 2.15 How a phagocyte ingests a pathogen

Exam tip

There is no need to be over-dramatic in answers about the immune system. It is not genocide. Words such as 'kill' and 'destroy' should be used sparingly. 'Annihilate' is definitely over the top.

Exam tip

Phagocytes don't 'eat' bacteria; they engulf and digest them. 'Eating' should only be used when describing animals that have a mouth and a gut.

The response of T lymphocytes to a foreign antigen

REVISED

When a particular pathogen gets into the body, the T lymphocytes respond as follows. This is known as the **cellular response**.

1 The pathogen is recognised as foreign and engulfed by a phagocyte.
2 This cell takes the antigens from the pathogen and 'displays' them on its membrane. It becomes an **antigen-presenting cell**.
3 The antigens are detected by helper T cells, which become activated.
4 In turn, the helper T cells activate three other types of cells:
   - phagocytes, which carry out phagocytosis (see above)
   - cytotoxic T cells (TC cells), which carry out cell-mediated immunity (see below)
   - B lymphocytes, which carry out humoral immunity (see below)

Cell-mediated and humoral immunity

REVISED

The immune response takes two different forms: **cell-mediated immunity** and **humoral immunity**.

The cytotoxic T cells carry out cell-mediated immunity, meaning that the cell itself attacks the pathogen. TC cells attach to the pathogen and inject toxins into it, causing its death, or label the pathogen for phagocytosis. TC cells can also become **memory cells**.

The humoral response involves B lymphocytes which, when exposed to an antigen, form plasma cells. These cells produce and secrete antibodies to a specific antigen. A small number remain as memory cells. If cells carrying the same antigen enter the blood again, the memory cells recognise them and produce new plasma cells faster than before. A humour is a fluid, so humoral immunity means 'done by the fluid' and it refers to the fact that there are antibodies in the plasma (i.e. fluid).

Passive and active immunity

REVISED

If the body is to survive exposure to a particular disease, it needs to be able to make antibodies in sufficient quantities to neutralise the pathogen. The problem is, we can only make enough antibodies if we have already been exposed to the pathogen. How do we survive that first exposure?

An unborn baby grows and develops in a sterile environment. Before birth, it gets antibodies across the placenta so that when it is born it already has some immunity passed on by its mother. After birth, the baby continues to acquire some immunity from its mother via her milk. Getting antibodies ready-made from the mother is called **passive immunity**. When the body makes its own antibodies, this is known as **active immunity**.

The primary and secondary immune responses

REVISED

Primary response

As soon as we are born, we are exposed to a variety of pathogens. The immune system develops by **clonal selection:** at birth, we have millions of different types of B lymphocyte, each capable of making a particular antibody.

When a particular pathogen gets into the body for the first time, the T lymphocytes activate the B lymphocytes capable of making the right antibody, as described above. The B lymphocytes multiply into a large population of plasma cells and memory cells. This **primary immune response** generally takes too long to be effective in preventing symptoms, but it does create long-lasting memory cells.

Secondary response

The **secondary immune response** is the one you want. If we already have memory cells in place from the first exposure, a second exposure stimulates the memory cells to multiply rapidly into a clone of plasma cells, which can make the right antibody quickly enough to prevent symptoms developing (Figure 2.16). We are immune to a disease as long as we have the memory cells in place and can therefore produce the secondary immune response when exposed to a particular pathogen.

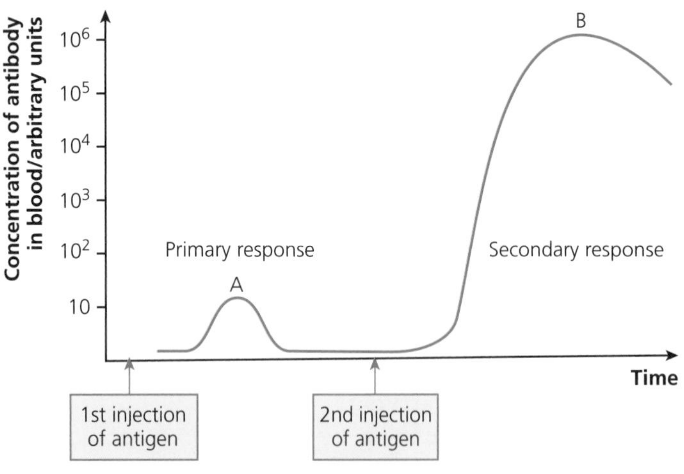

Figure 2.16 The primary and secondary immune responses

Exam tip

The *y*-axis uses a logarithmic scale, which allows a large range of values to be shown.

Now test yourself

TESTED

16 Suggest why breast milk is considered more preferable for young babies than powdered formula milk.
17 Explain what is meant by the term *memory cell*.
18 Outline the differences between the primary immune response and the secondary immune response.

Answer on p. 204

Antigenic variability in the flu virus

Antigenic variation refers to the mechanism by which an infectious organism such as a bacterium or virus alters its surface proteins in order to evade a host's immune response. The flu virus is well known for this ability: it rapidly mutates so that new strains, with new antigens, are constantly being produced. Any antibodies made against an old strain will not work against a new one.

Vaccines

REVISED

There are several types of **vaccination** that are used to provide protection for individuals and populations against disease:

- dead pathogens
- live but attenuated pathogens (ones that have been treated so that they cannot cause disease)
- purified antigens

Exam tip

The terms *vaccination* and *immunisation* mean the same thing.

Whatever the type of vaccine, they all contain antigens and stimulate the primary response so that the body has memory cells in place. If or when the actual pathogen enters the body, the memory cells multiply into plasma cells, which in turn produce antibodies in large enough quantities to fight the infection before symptoms appear.

Herd immunity

Herd immunity is the idea that you do not have to vaccinate everyone in order to stop an epidemic. You just have to vaccinate a large enough proportion of the population in order to break the chain of transmission, providing a measure of protection for individuals who are not immune.

HIV and AIDS

REVISED

The structure of the **human immunodeficiency virus (HIV)** is shown in Figure 2.17. It is classed as a retrovirus because its genetic material is RNA. In order to reproduce, the virus must turn its RNA into DNA, transcription in reverse, so it possesses the enzyme **reverse transcriptase**.

HIV causes the symptoms of **AIDS** because it invades and reproduces inside a particular type of helper T cell, called a CD4 cell, killing them or preventing their replication. If left untreated, HIV infection reduces the helper T cell population to a level where they no longer activate other cells and the immune system is no longer effective. The body is left vulnerable to opportunistic infections and death is caused by pathogens that would normally cause little trouble.

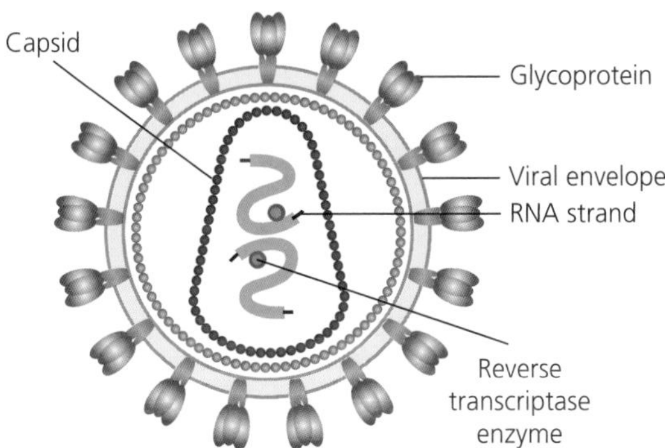

Figure 2.17 The HIV virus

Antibiotics work because they interfere with some aspect of prokaryote cell function while not harming eukaryote cells. Antibiotics do not work against viral diseases such as HIV because viruses do not have internal organelles or any metabolism that can be targeted.

Monoclonal antibodies

REVISED

Monoclonal antibodies are produced by B cells in culture. They are useful in science and medicine because they are specific and bond with just one type of substance. They can detect minute quantities of a particular substance and are therefore important in **medical diagnosis**. For example, in pregnancy tests, antibodies detect very small quantities of the hormone hCG. The more sensitive the test, the earlier it can be done.

The problem is that large quantities of antibody are needed, but B lymphocytes do not grow well in culture outside the body. If B lymphocytes are joined with certain cancer cells, the resulting cells will divide again and again, while making useful amounts of the desired antibody.

In addition to diagnosis, monoclonal antibodies can be used as therapeutic agents because of their ability to seek out particular cells. For example, certain tumours may express a particular protein on their cell-surface membrane. If an anti-cancer drug is attached to the antibody complementary to these proteins, it will seek out and deliver the drug to exactly the cancerous cells that need to be targeted.

Ethical issues with vaccines and monoclonal antibodies

REVISED

The use of vaccines seems pretty uncontroversial, but there are issues:

- They may be produced using animal tissue such as eggs, so strict vegans may object to receiving them.
- In a pandemic (widespread outbreak), when vaccinations are urgent, who has priority?
- Should people have the right to say no to a vaccination? It is currently unlawful to force anyone to have a vaccination but, in refusing, is an individual putting others at risk?

The issues with monoclonal antibodies include:

- They are produced using live animals, usually mice. There is a need to induce tumours in the mice, which some people see as unacceptable.
- Monoclonal antibodies have been used successfully to treat life-threatening diseases, including some types of cancer. However, a number of trials involving the use of monoclonal antibodies to treat disease have been fatal.

The ELISA test

REVISED

The **ELISA test** (enzyme-linked immunosorbent assay test) is a sensitive test that uses monoclonal antibodies to detect particular substances, often antigens. The test usually involves particular enzymes and their substrates, an antigen–antibody reaction and a visible colour change.

Now test yourself

TESTED

19 Use your knowledge of herd immunity to explain why some people think that vaccinations should be made compulsory for all.

Answer on p. 204

Exam practice

1 The graph shows the results of nine cases of breast cancer in which a new combined chemotherapy and microwave treatment is compared with chemotherapy alone. The aim is to shrink the tumour so that it can be removed effectively by surgery without having to remove the whole breast.

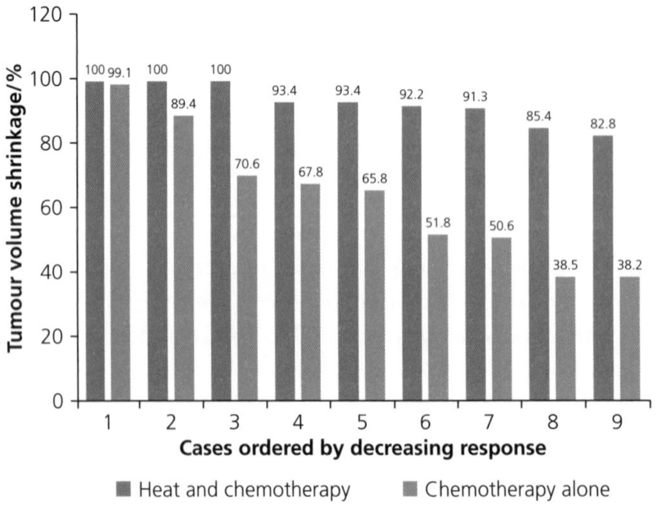

Source: www.ll.mit.edu/news/clinicaltrial.html

(a) What is meant by a tumour volume shrinkage of 100%? [1]
(b) In case 3, the breast tumour had an initial volume of 150 cm^3 (estimated by ultrasound). What was its volume after chemotherapy alone? [1]
(c) Give two conclusions supported by the data. [2]
(d) Evaluate the conclusion that heat and chemotherapy treatment should be used in all cases of breast cancer. [3]

2 Antibodies are proteins. Explain how the structure of proteins allows them to form many different types of antibody. [2]
3 Explain the importance of antigens in the process of phagocytosis. [2]
4 Vulnerable people are offered a flu vaccine every winter.
(a) Suggest what is meant by the term *vulnerable*. [1]
(b) Explain why a new vaccine is needed each year. [2]
5 Use your knowledge of the primary and secondary immune responses to explain how vaccines protect people from disease. [2]
6 Monoclonal antibodies are used in pregnancy tests. Explain how the properties of antibodies make them useful in diagnostic tests. [2]
7 Suggest why a parent might object to their child receiving a particular vaccination. [1]

Answers and quick quiz 2 online

ONLINE

Summary

By the end of this chapter you should be able to understand:

- The structure and function of the main organelles in animal and plant cells.
- The structure and function of the main organelles in prokaryotic cells.
- The differences between eukaryotic and prokaryotic cells.
- The essential features of a virus.
- The principles of and differences between optical and electron microscopes.
- How to do calculations involving actual size, observed size and magnification.
- The principles of cell fractionation (how to get samples of pure organelles).
- The stages involved in mitosis.
- Interphase and the cell cycle.
- The relationship between the cell cycle and cancer, and its treatments.
- How to calculate a mitotic index.
- The basic structure of all cell membranes (the fluid-mosaic model of membrane structure).
- How substances move across cell membranes.
- The process of phagocytosis, including the role of lysosomes.
- The definitions of antigen and antibody.
- Antibody structure and the formation of an antigen–antibody complex.
- The cellular and humoral responses and the role of B lymphocytes and T lymphocytes.
- The difference between the primary and secondary immune response and the role of plasma cells and memory cells.
- Why the flu virus is so variable and the problems this causes.
- How vaccines provide protection against disease.
- What is meant by *herd immunity*.
- How HIV causes AIDS.
- Monoclonal antibodies and their uses.

3 Organisms exchange substances with their environment

Surface area to volume ratio

The relationship between the size of an organism and its surface area to volume ratio

REVISED

Size matters a lot in biology. All organisms need to exchange material with their surroundings. Generally, they need nutrients and oxygen, and need to get rid of carbon dioxide and other wastes. There are two basic rules:

- the amount of material an organism *needs* to exchange depends on its volume
- the amount of material it *is able* to exchange depends on its surface area

As an organism increases in size, its volume increases, it has more cells and so it needs more from its environment. However, its surface area does not increase as quickly as its volume, so the larger an organism gets, the more difficult it becomes to absorb enough substances over its outer surface. Table 3.1 and Figure 3.1 show what happens to surface area, volume and **surface area to volume ratio** as an organism gets larger.

The **surface area to volume ratio (SA : V)** is the surface area of an organism divided by its volume. It is a key concept as the surface area must be able to provide sufficient oxygen through diffusion from the environment.

Table 3.1 The change in surface area and volume as an organism gets larger

Length of organism (l) (mm)	Surface area of organism ($6 \times l^2$) (mm^2)	Volume of organism (l^3) (mm^3)	Surface area to volume ratio
1	6	1	6 : 1
2	24	8	3 : 1
3	54	27	2 : 1
10	600	1000	0.6 : 1

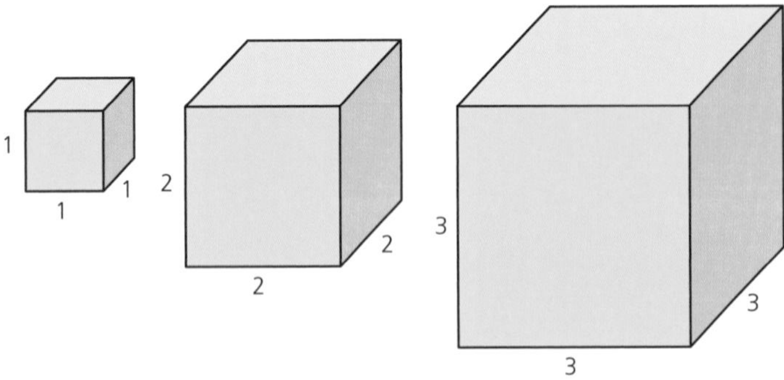

Figure 3.1 As the volume of a cube increases, its surface area to volume ratio decreases

So if an organism is large, it needs to develop adaptations that increase its surface area. There are two ways of doing this:

- having a flat or cylindrical (worm-shaped) body. This increases the surface area to volume ratio, so that all cells are a short diffusing distance from the outside
- having organs that increase the surface area, such as gills. This usually means that the organism also needs some sort of circulatory system to distribute the oxygen to the other parts of the body

Exam tip

Candidates often fail to appreciate that although large animals such as elephants must have a large surface area, they have a very small surface area compared with their volume.

Size and metabolic rate

REVISED

Metabolism is the general term for the biochemical reactions that occur inside an organism and **metabolic rate** is the speed of these reactions. In practice, metabolic rate is the same as the respiration rate and it can be determined by measuring the oxygen used by an organism.

Warm-blooded animals such as mammal and birds maintain a constant core body temperature, so they must balance heat generated with heat lost. Heat is generated by respiring cells and so the amount of heat made depends on an organism's volume. However, the amount of heat lost varies according to surface area. The smaller the animal, the higher its metabolic rate. This is because:

smaller animals have a high surface area to volume ratio

so they lose heat quickly to the surroundings

and have to respire quickly to generate heat to replace what is lost

so they must consume more food and oxygen per unit of body weight than larger animals

With a small mammal like a shrew, 99% of the food it eats is respired to maintain its body temperature. This is why it must consume almost its own body weight in food every day and its heart rate is over 800 beats per minute.

Now test yourself

TESTED

1 Suggest suitable units for measuring metabolic rate. Remember that you need to be able to compare organisms of different sizes.

Answer on p. 204

Gas exchange

Adaptations of gas exchange surfaces

REVISED

Why do organisms need to exchange gas? It is all about **respiration** — the process that releases the energy locked in organic molecules such as glucose and lipids. All cells, in all living things, respire all the time. The process requires a constant supply of oxygen and produces carbon dioxide that needs to be eliminated. In addition, photosynthesising tissues need to absorb carbon dioxide and remove waste oxygen.

Exam tip

Ventilation and breathing refer to the same process, but respiration is different. Make sure you use the correct terms.

Single-celled organisms

Single-celled organisms are microscopic and have a very large surface area to volume ratio. They can exchange gas over their whole body surface. Gas exchange is rapid because the diffusing distances involved are very small.

Insects

If you look closely at an insect, you may see series of holes called **spiracles**, which are arranged in a row along the side of the body (Figure 3.2). Spiracles lead to breathing tubes (**tracheae**) that take air directly to the respiring tissues. The trachea branch out into a fine network of **tracheoles** which are so small and dense that they pass close to — and sometimes go inside — respiring cells. Diffusion through air is so fast that oxygen can be delivered quickly enough to supply the cells.

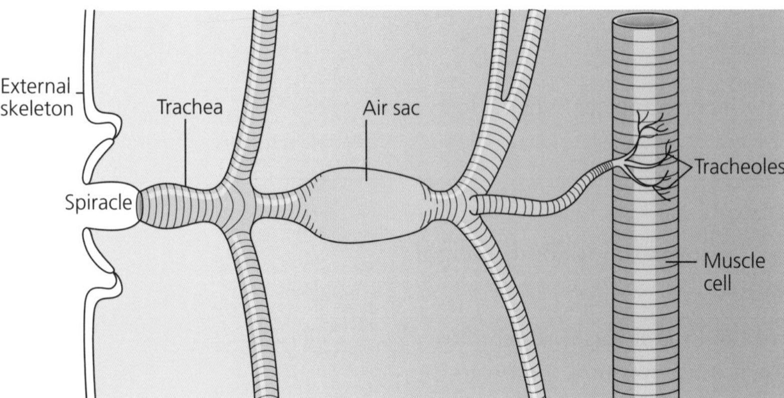

Figure 3.2 The tracheal system of an insect

Now test yourself

TESTED

2 Explain why single-celled organisms such as amoeba do not need organs of gas exchange.
3 The abdomen of an active wasp can be seen pulsing. Use Figure 3.2 to suggest a reason for these movements.

Answers on p. 204

Minimising water loss

Most insects are **terrestrial** — they live on land. There are a few aquatic insects in freshwater, but virtually none in the oceans. Minimising **water loss** is important. Insects have two key adaptations:

- they have an exoskeleton made from waterproof **chitin**
- the spiracles can close when oxygen demand is low. Some spiracles are surrounded by hairs that minimise air movement, and therefore water loss

Fish

It is not easy to breathe in water. Compared with air, water has much less oxygen but is about 800 times denser. It takes a lot of energy to move water. In addition, diffusion of oxygen and carbon dioxide is much slower through water than through air (Figure 3.3).

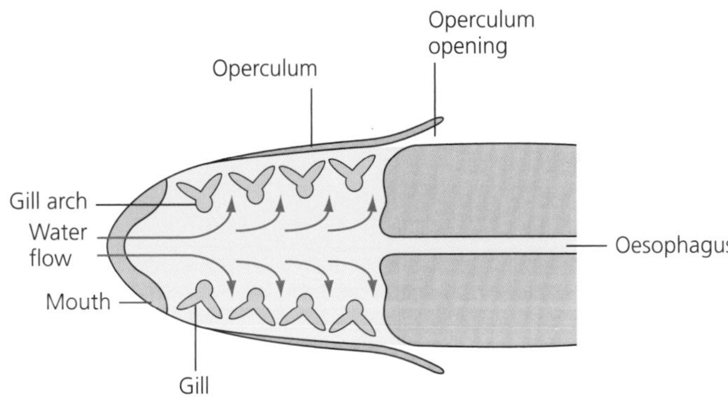

Figure 3.3 The flow of water over the gills of a fish

Gills are gas exchange organs that have evolved to overcome these problems. Instead of taking in water, stopping it and then forcing it out again — like we do with air — **fish** open their mouths and allow the water to flow in one direction, over the gills and out through the operculum (gill cover).

Gills have all the adaptations you would expect in gas exchange surfaces (Figure 3.4):

- Large surface area — each gill has several arches or **rakers** (typically 3 to 5) that support many **gill filaments**. On each filament are numerous **lamellae** — delicate flat structures that are the equivalent of alveoli.
- Small diffusing distance — the lamellae have very thin cells so there is a short diffusing pathway between water and blood.
- Efficient blood supply — a clever countercurrent system.

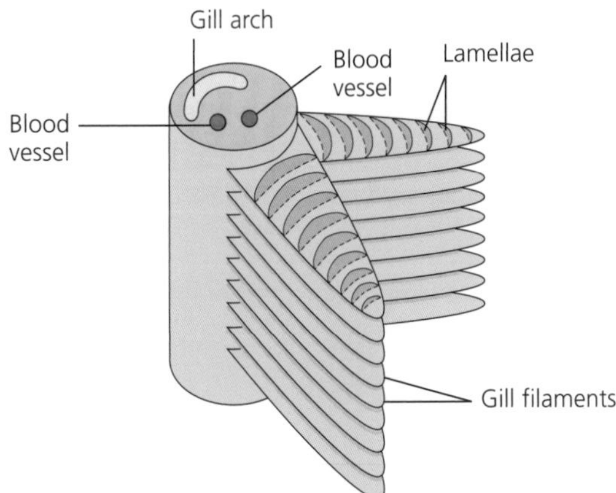

Figure 3.4 The basic arrangement of arches, filaments and lamellae

The countercurrent principle

The **countercurrent principle** involves two fluids flowing in opposite directions. This is an efficient way of maximising the exchange between the two fluids. In fish gills, the blood in the lamellae and the water flow in opposite directions. This means that there is always more oxygen in the water than in the blood. The key point to remember for exams is that this maintains a **diffusion gradient** (Figure 3.5).

(a)

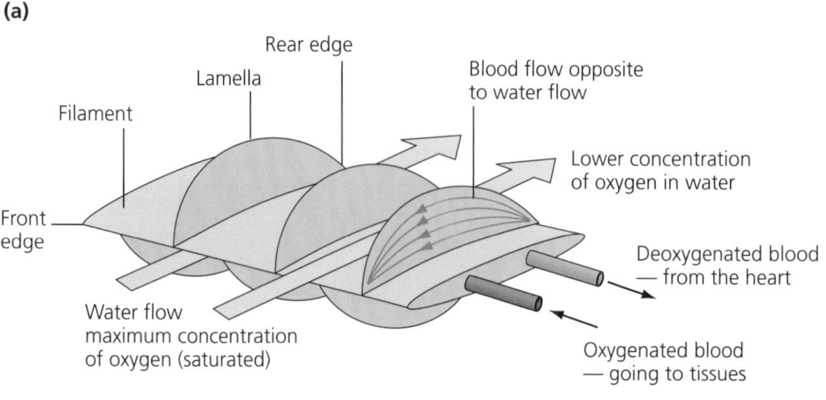

(b)

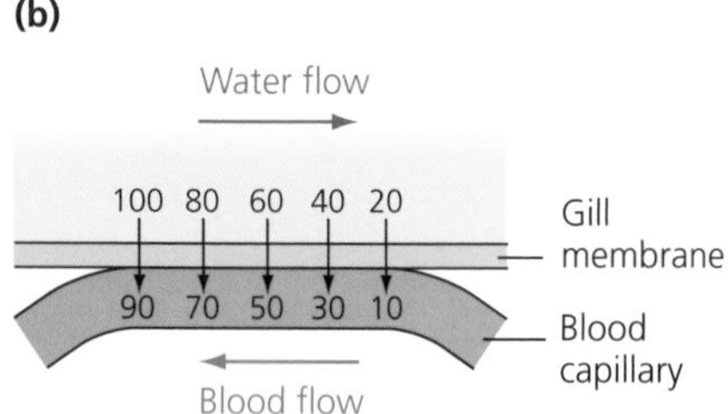

Figure 3.5 The countercurrent principle. (a) Water and blood flow in opposite directions. (b) Maintaining a diffusion gradient. The figures refer to the percentage saturation with oxygen

Now test yourself

TESTED

4 In fish gills, the blood in the capillaries flows in the opposite direction from the water passing over them. Explain the advantage of this system.
5 Use Figure 3.5 to predict what would happen if the water and blood were to flow in the same direction.

Answers on p. 204

Leaves

Plants need to exchange gas too. Leaves are organs of photosynthesis and, when it is light, the chloroplasts in the palisade cells need a supply of carbon dioxide and need to release the excess oxygen. Plant cells respire all the time, so when there is no light they need oxygen and make carbon dioxide. In order to maximise gas exchange, the leaves of a **dicot plant** have:

- loosely packed spongy mesophyll cells that create air spaces
- stomata — holes that allow direct access between the air spaces and the atmosphere

Exam tip

There are two types of flowering plant, monocots and dicots. Monocots are grasses and related species, white the dicots contain most other familiar groups.

Xerophytic plants are adapted to living in dry (or arid) places. The following adaptations help them to reduce water loss:

- thick waxy cuticle on the leaves
- smaller leaf area
- stomata in pits
- hairy leaves
- rolled leaves

The human gas exchange system

REVISED

The **lungs** are organs that are adapted for gas exchange in air. Their function is simple: to get as much fresh air as possible in contact with blood. As you can see from Figure 3.6, the bronchial tree is a branching system of tubes that takes air to the **alveoli** deep in the lungs. All of the tubes are held open by rings of cartilage except the **terminal bronchioles**.

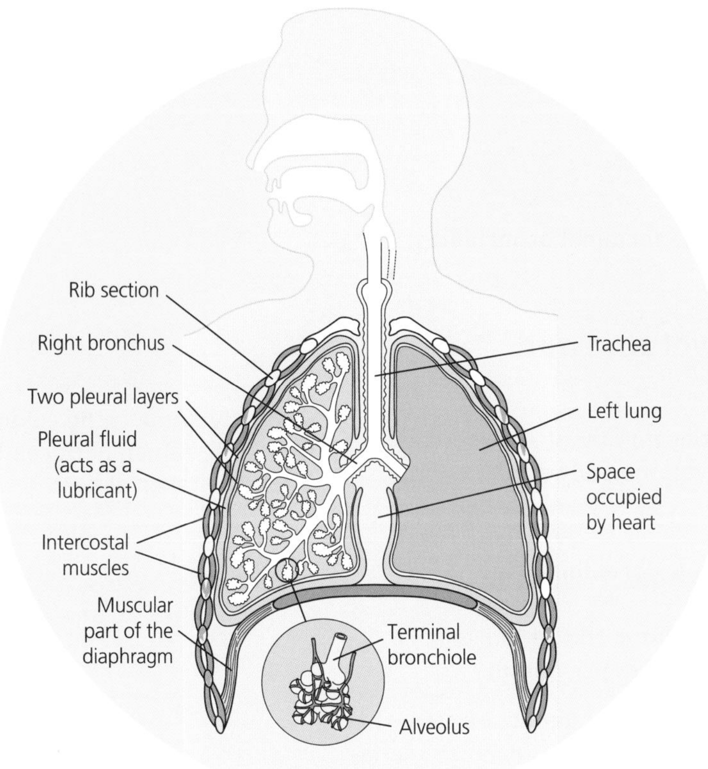

Figure 3.6 The structure of the human gas exchange system

The alveolar epithelium

The **alveolar epithelium** is the surface over which gas exchange takes place. Its structure is related to function:

- Millions of alveoli provide a large surface area.
- The alveolar walls consist of flat cells called **squamous epithelium** that are as thin as possible. This makes the diffusing pathway as small as possible.
- The dense network of blood capillaries around the alveoli means that a lot of blood is in close contact with the air.
- The beating of the heart, along with constant breathing, makes sure that deoxygenated blood meets fresh air. This maintains a diffusion gradient.

All these features serve to make diffusion as fast and efficient as possible. The rate of diffusion of a substance across an exchange surface is inversely proportional to the thickness of the exchange surface.

The exchange of gases at the alveoli

The exchange of gases at the alveoli occurs by simple diffusion (Figure 3.7). Oxygen and carbon dioxide are small, simple molecules that pass easily through cell walls. There is more oxygen in the air than in the blood, so oxygen diffuses into the blood. There is more carbon dioxide in the blood than in the air, so this gas diffuses in the opposite direction.

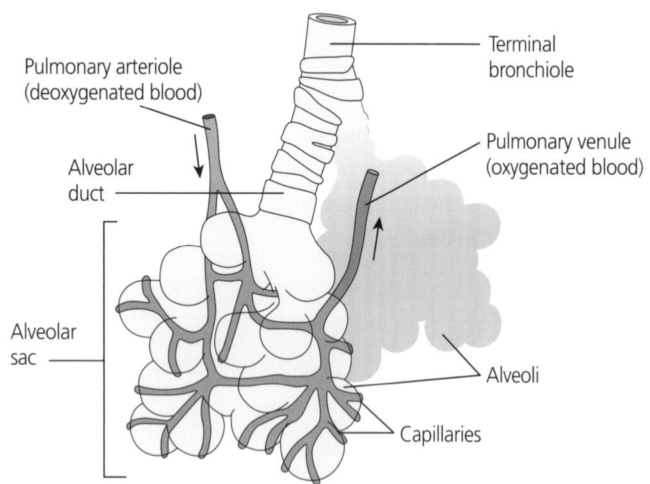

Figure 3.7 **Alveoli are found at the end of the terminal bronchioles**

> **Exam tip**
>
> Gas exchange always occurs by diffusion, not by active transport, facilitated diffusion or osmosis.

Pulmonary ventilation, tidal volume and ventilation rate

Pulmonary ventilation rate (PVR) is the amount of air we inhale in 1 minute. It is calculated as using **tidal volume** and ventilation rate, as follows:

> **Tidal volume** is the volume of air breathed per breath, usually taken at rest.

pulmonary ventilation rate = tidal volume × breathing rate

Typical values would be 0.5 litres for tidal volume and 14 breaths per minute, giving a pulmonary ventilation of 7 litres per minute.

Lung volumes are measured using a spirometer, which produces a trace like the one in Figure 3.8. When we exercise, our breathing undergoes a series of changes. Key points on the trace are labelled with the letters **A** to **E**:

- **A** to **B** shows normal, tidal breathing while at rest.
- **B** to **C** shows the effect of exercise — breathing is deeper and more frequent.
- **C** to **D** shows a return to normal after exercise.
- **D** shows the subject breathing out as far as possible.
- **E** shows the subject inhaling as much as possible.
- The difference in volume between **D** and **E** shows the vital capacity — the total usable lung volume
- The residual volume is the air that must remain in the lungs.

> **Exam tip**
>
> You may have been taught terms such as inspiratory reserve volume and expiratory reserve volume, but they are not in the specification and so cannot be tested in the exam.

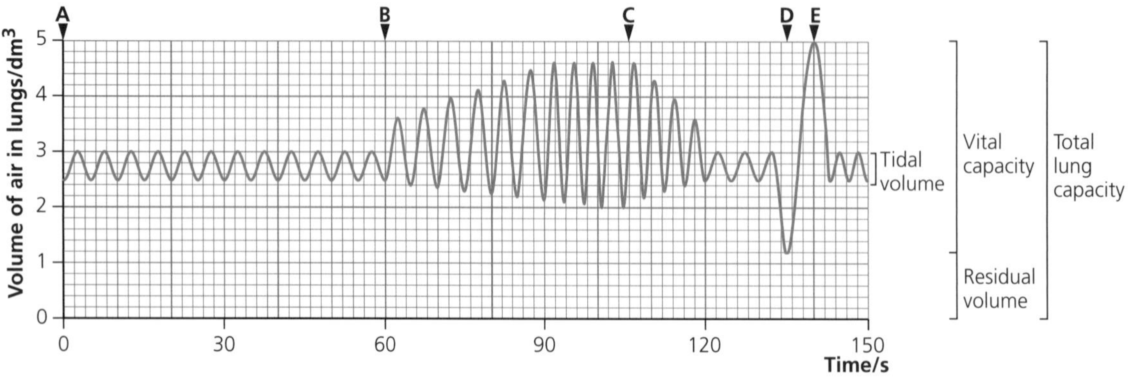

Figure 3.8 **Changes in breathing during exercise**

The mechanism of breathing

The lungs are spongy organs that simply fill with air and then empty again. They cannot do this on their own because they have no muscle. To inflate, they need the **intercostal muscles** and the **diaphragm**.

The **mechanism of breathing** is shown in Table 3.2. Inspiration is an active process — it requires energy. Expiration is largely a passive process. The lungs, diaphragm and intercostal muscles are all elastic. When stretched, they tend to go back to their original shape.

Table 3.2 The mechanism of breathing

Stage	Breathing in (inspiration)	Breathing out (expiration)
1	The external intercostal muscles contract, pulling the ribs up and out	The external intercostal muscles relax
2	The diaphragm muscles contract, flattening the diaphragm	The diaphragm relaxes. The abdominal organs (liver and intestines) push upwards
3	Their combined effect is to increase the volume of the thoracic cavity	The volume of the thoracic cavity decreases
4	This lowers the pressure in the thoracic cavity to below atmospheric pressure	The pressure inside the thoracic cavity increases
5	Atmospheric pressure forces air into the lungs	Air is forced out

The biological basis of lung disease

REVISED

The AQA specification does not list any particular lung diseases that need to be learnt in detail. However, it does say that students need to be able to interpret the effects of disease on gas exchange and/or ventilation, so it helps to know about the some common lung problems.

Emphysema

Fibrosis is a response to tissue damage. When normal body tissue such as lung, liver or heart becomes damaged, the body responds by producing connective tissue, commonly called scar tissue. Emphysema is fibrosis of the alveoli. Lung damage results from long-term exposure to irritants such as cigarette smoke and air pollution (Figure 3.9). When the delicate alveolar walls become damaged and replaced by connective tissue, the effects are:

- reduced surface area
- increased diffusing pathway due to the thicker walls
- decreased elasticity, so that the lungs cannot expand as much and it takes more effort to breathe out

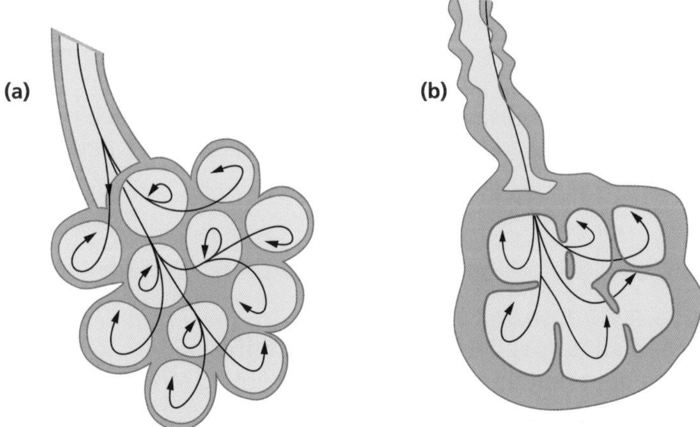

Figure 3.9 (a) Healthy alveoli. (b) Alveoli in a person with emphysema

All three effects combine to make gas exchange less efficient. Emphysema is long-term, irreversible damage to the lungs.

Emphysema often occurs in people who also have bronchitis, which is a condition that makes bronchiole walls inflamed and covered with more mucus than normal. Together, these two conditions are called chronic obstructive pulmonary disease (COPD).

Asthma

Asthma is a condition in which the muscles lining the terminal bronchioles constrict. The lining also over-produces mucus. Both of these responses narrow the airway, making it particularly difficult to breathe out. The narrow airway produces a wheezing sound during exhaling. In asthmatics, an attack is brought on by an environmental trigger such as stress or cold air, although it is usually triggered by an allergen such as house dust or animal fur. Asthma affects lung function because it reduces air flow. It does not damage alveoli and so there is no reduction in surface area.

Example

Graphs and patterns

Study the graph, which shows the change in the number of asthma patients and four common air pollutants over a 20-year period. The air pollutants are particle matter, sulfur dioxide, nitrogen oxide and carbon monoxide.

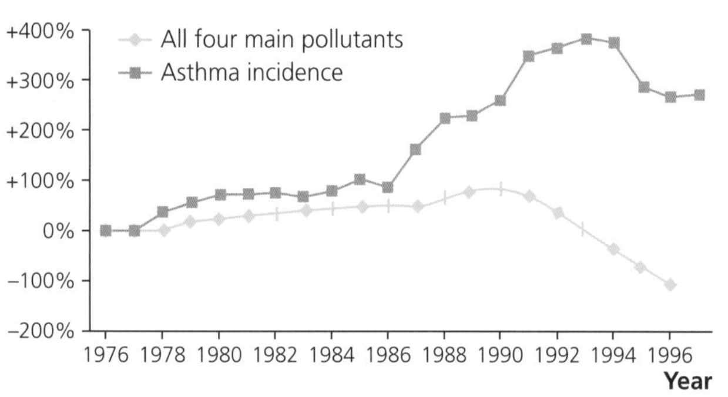

(a) Describe the pattern of incidence of asthma. [3]
(b) Evaluate the claim that air pollution causes asthma. [3]

Answer

You have to match up your answer to the number of marks. If the graph is one straight line, there will usually be 1 mark for writing something like 'as X increases, Y increases'. However, for a more complex graph like the one here, there will often be 2 or 3 marks available.

(a) There is a slow increase up to 1986. There is a rapid increase from 1986 to 1993. There is a fall after the peak in 1993.
(b) There is a correlation between the number of pollutants and the incidence of asthma. However, a correlation does not mean a cause. Something else could be causing the increase, such as more allergens.

Exam tip

There will always be questions about data in the exam. Tables and graphs are nothing to be afraid of. In fact, they are often easy marks as the information is there for you.

Exam tip

Part (a) asks you to 'describe', so there is no need to explain anything here. Part (b) asks you to 'evaluate', which means to look at both sides.

Now test yourself

TESTED

6 Explain the difference between the terms *respiration* and *breathing*.
7 List three ways in which the lungs are adapted to speed up the process of diffusion.
8 Explain why people with emphysema often get breathless when they do anything strenuous.

Answers on p. 204

Digestion and absorption

The digestive system

REVISED

The intestine (or gut) is a long tube that passes through the middle of the body (Figure 3.10). Food goes in the mouth and is digested so that it can be absorbed into the blood. Anything that cannot be digested cannot be absorbed and so passes straight through. The content of the gut is not part of your body. It is a mixture of partially digested food, cells from the gut lining, digestive enzymes and bacteria.

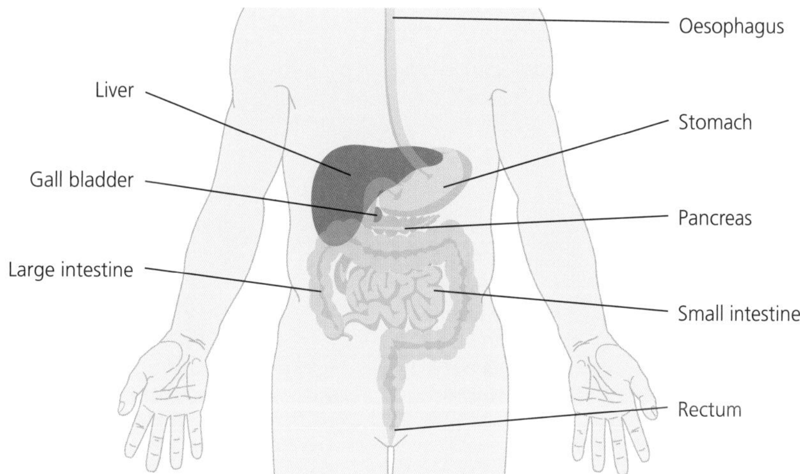

Figure 3.10 The human digestive system

The molecules in our food are needed to build and maintain our bodies. The three key groups of organic molecules are carbohydrates, lipids and proteins. The food we eat consists of a complex mixture of these three types, together with simpler ones such as water, vitamins and salts.

Digestion involves breaking down these large molecules so that they are simple, soluble and can be absorbed into the blood. **Large biological molecules** are **hydrolysed** by enzymes to produce **smaller molecules** that can be absorbed across **cell membranes**. Once inside the body, smaller molecules are built up into large ones by condensation.

Exam tip

Condensation and hydrolysis are common themes for exam questions. Make sure you get them the right way round. Condensation reactions *produce* water, whereas hydrolysis reactions *use* water.

Now test yourself

TESTED

9 Explain the difference between condensation and hydrolysis reactions.

Answer on p. 204

Digestion

REVISED

Carbohydrate digestion

Carbohydrate digestion begins in the mouth. Saliva contains the enzyme **salivary amylase**, which hydrolyses starch into maltose. This digestion is not really significant because:

- we generally swallow food before the enzyme has a chance to work
- hot food can denature the enzyme
- some people do not make salivary amylase — it is genetic

There is no carbohydrate digestion in the stomach. The acidic pH in the stomach stops the salivary amylase from working and the stomach enzymes digest only protein.

The main region for carbohydrate digestion is the small intestine. Food leaving the stomach and entering the small intestine has two digestive juices added to it: pancreatic juice from the pancreas and bile from the liver. Bile has no part in carbohydrate digestion but pancreatic juice is vital.

1 Starch is digested by **pancreatic amylase** in pancreatic juice. This produces maltose, which joins lactose and sucrose already in the gut. So there are three disaccharides that need digesting.
2 The disaccharides are hydrolysed into monosaccharides by **maltase**, **sucrase** and **lactase**. These enzymes are fixed in the membranes of the intestinal epithelial cells (Figure 3.11). The monosaccharides can then be absorbed.

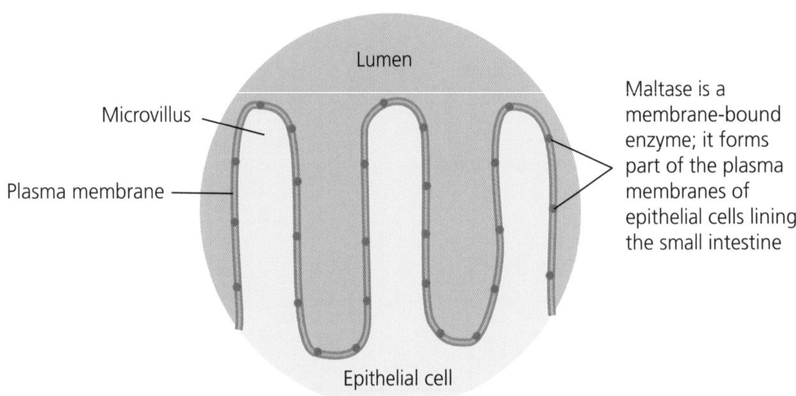

Figure 3.11 The location of the maltase enzymes in the wall of the intestine

Now test yourself

TESTED

10 Where are the enzymes that break down maltose located?
11 Name the main region of the gut where carbohydrate digestion takes place.
12 Why is it an advantage to have enzymes fixed in the wall of the intestine rather than having them free in the lumen?

Answers on p. 204

Lipid digestion

Lipids and water do not mix and so any lipids eaten tend to reach the small intestine in globule/droplet form. **Lipid digestion** is a two-part process:

1 Physical digestion. **Bile salts** in bile lower the surface tension between lipids and water so that large droplets are split into many smaller ones. This process is celled **emulsification** and it greatly increases the surface area of the lipid droplets.
2 Chemical digestion. The enzyme **lipase**, secreted in pancreatic juice, hydrolyses triglycerides into fatty acids and glycerol.

The end result of these processes is a mixture of glycerol, fatty acids and monoglycerides (glycerol attached to one fatty acid). The bile salts form **micelles**, which are tiny droplets (4–8 nm across) that contain the products of lipid digestion on their surface. Micelles are constantly ferrying the components of lipid digestion from the lumen to the epithelial cells so that they can be absorbed.

Now test yourself

TESTED

13 Lipase enzymes are water soluble, so they only work on the surface of lipid droplets. Use this information to explain why bile salts are so important.

Answer on p. 204

Protein digestion

Protein digestion takes place in both the stomach and the small intestine. There are several different **protease** enzymes, also called **proteolytic** enzymes, but they all work by hydrolysing peptide bonds. They can be roughly divided into three types:

- **endopeptidases**, which cut within the protein, turning long chains of polypeptides into shorter chains
- **exopeptidases**, which remove the terminal (end) amino acids from polypeptide chains
- **dipeptidases**, which are membrane-bound enzymes similar to the enzymes that digest disaccharides. They hydrolyse dipeptides into individual amino acids

It is important to appreciate that these enzymes combine to be far more effective than each single type would be on its own. For example, endopeptidase enzymes make more 'ends' for the exopeptidase enzymes to work on.

Absorption

REVISED

Adaptations of the ileum

The mammalian **ileum** (the main part of the small intestine) is adapted to absorb the products of digestion as follows (Figure 3.12):

- It is long.
- Its lining is folded into millions of projections called villi.
- The epithelial cells covering the villi are themselves folded into many microvilli.

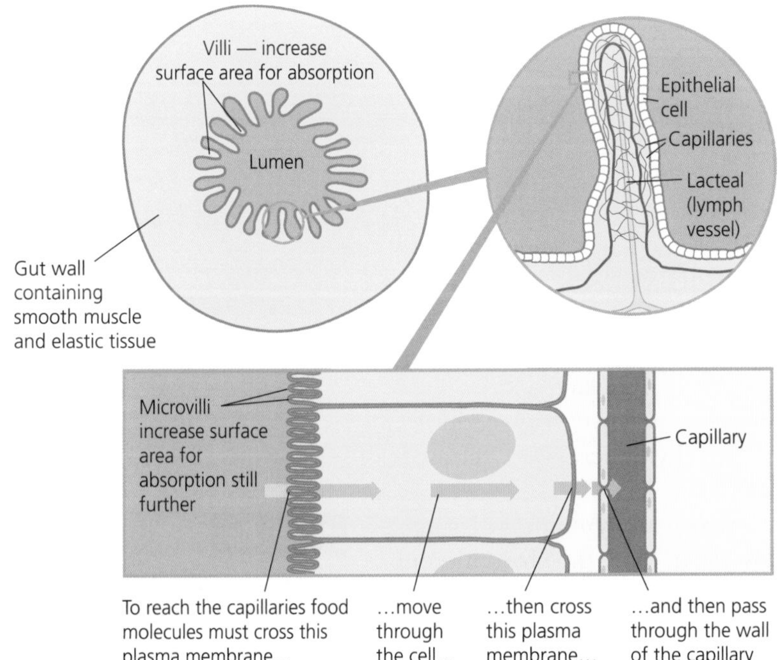

Figure 3.12 How digested food molecules reach the capillaries in the gut wall

The above adaptations all combine to create a massive surface area of membrane in direct contact with the gut contents. There are lots of digestive enzymes and transport proteins embedded in the membrane. In addition:

- the epithelial cells themselves are thin, creating a short diffusing pathway
- the epithelial cells have many mitochondria, to synthesise ATP for active transport
- there is an efficient blood supply running through all the villi, so that substances are transported away as soon as they are absorbed. This speeds up absorption by maintaining a diffusion gradient

Amino acids

Amino acids are absorbed by a process known as **co-transport**. The membranes of the epithelial cells contain several different amino acid transporters, which absorb the different types of amino acid such as acidic, basic or neutral. They all transport amino acids only after binding with sodium. Once a symport protein is loaded up with both an amino acid and sodium, it undergoes a shape change that transports both molecules to the other side before going back to its original shape. The whole process is dependent on the concentration gradient of sodium. If there is no sodium gradient, there is no amino acid absorption. The whole process is almost identical to the co-transport of glucose — see page 44.

Monosaccharides

Once the digestive enzymes have converted all the carbohydrates in the diet, the resulting **monosaccharides** need to be absorbed. By far the most common monosaccharide is glucose, which (like amino acids) is absorbed by co-transport. It involves both facilitated diffusion and active transport, and absorbs sodium ions (Na^+) at the same time. See p. 44 for more details on the absorption of glucose.

Exam tip

You are expected to know the cellular and molecular details of absorption. Make sure you can explain how the fine structure of an intestinal epithelial cell adapts it for the absorption of digested food.

Now test yourself

TESTED

14 The absorption of amino acids, glucose and sodium also causes the absorption of water from the gut. Explain how.

Answer on p. 204

Lipids

The absorption of lipids is a two-stage process:

1 The epithelial cells absorb the fatty acids, glycerol and monoglycerides because they can pass easily through the lipid part of the membrane. The micelles are not absorbed. Once inside the cell, all the components of lipid digestion pass to the endoplasmic reticulum there they are recombined into triglycerides. They are then packaged into **chylomicrons**, which contain, in order of abundance, triglycerides, phospholipids, cholesterol and protein. The chylomicrons are released from the epithelial cells by exocytosis.

2 The chylomicrons pass into the lacteals, which are lymph capillaries. There is one in the centre of each villus. From here, the chylomicrons are carried in the lymphatic system, finally joining the blood system in the subclavian vein, which is in the chest cavity, just under the collar bone. From here the chylomicrons are distributed around the rest of the body. Lipids are insoluble in water and so chylomicrons are the main method of lipid transport in the blood.

Exam tip

Whenever you are about to write the word *digested*, think about using *hydrolysed* instead. It makes examiners happy.

Mass transport in animals

When multicellular organisms develop organs of exchange such as lungs, guts and gills, they need a transport system to move substances over large distances because diffusion is simply too slow. Most transport systems consist of a series of tubes in which an efficient supply of materials is moved around under pressure. These systems are called **mass transport** systems. Plants have xylem and phloem, whereas vertebrates have a blood system.

Haemoglobin

REVISED

Red blood cells are unique cells that carry oxygen. They also, indirectly, help to carry carbon dioxide. The scientific name for red blood cells is erythrocytes, which just means 'red cells'. They are made in the bone marrow and are bi-convex in shape, which allows them to carry a useful amount of oxygen but to load and unload it quickly (Figure 3.13).

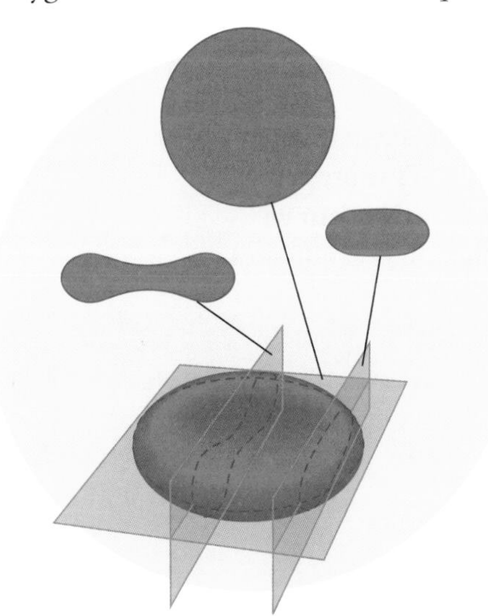

Figure 3.13 Red cells are bi-convex in shape, which is a perfect compromise between the maximum volume of a sphere and the maximum surface area of a flat disc

Now test yourself

TESTED

15 Red blood cells have no nucleus. Suggest why this is an advantage.

Answer on p. 204

Haemoglobin is a complex **protein** with a **quaternary structure**. It consists of four polypeptide chains (**globins**) attached to four iron-containing parts (**haems**) (Figure 3.14). At the centre of each haem is an Fe^{2+} ion that attaches to an oxygen molecule (O_2). As a chemical equation:

$$Hb + 4O_2 \rightleftharpoons HbO_8$$

Each haemoglobin molecule can combine with four oxygen molecules to form bright-red **oxyhaemoglobin**. The clever thing about haemoglobin is the cooperative nature of oxygen binding. When the first oxygen molecule combines with a haem group, the shape of the molecule changes so that it becomes easier for the other oxygen molecules to bind.

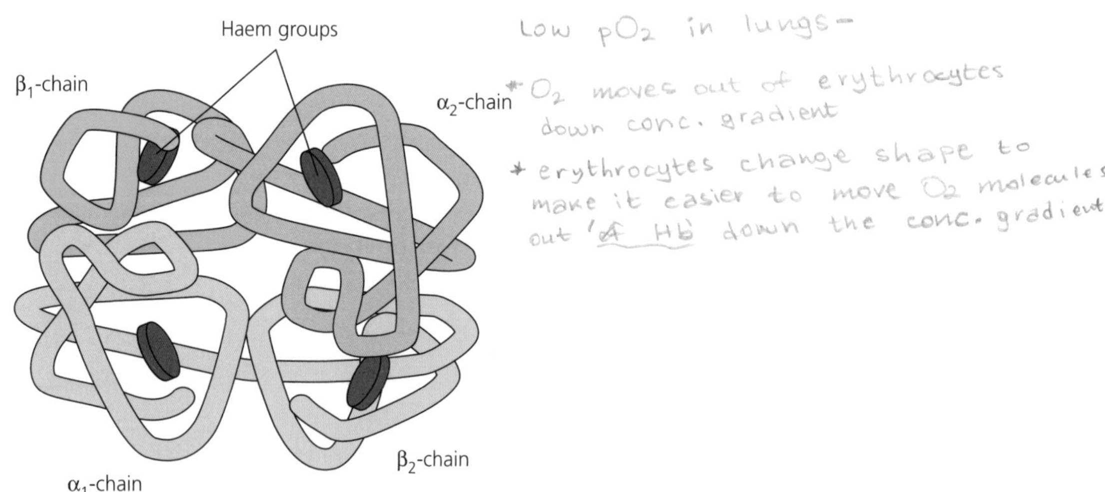

Figure 3.14 Haemoglobin is composed of four polypeptide chains, each attached to a haem group

The oxyhaemoglobin dissociation curve

The key property of haemoglobin is that it combines with oxygen where it is abundant, but then releases it again where it is scarce. This property is shown by a graph called the **oxyhaemoglobin dissociation curve** (Figure 3.15). This graph is a favourite topic of examiners, but candidates often find it difficult.

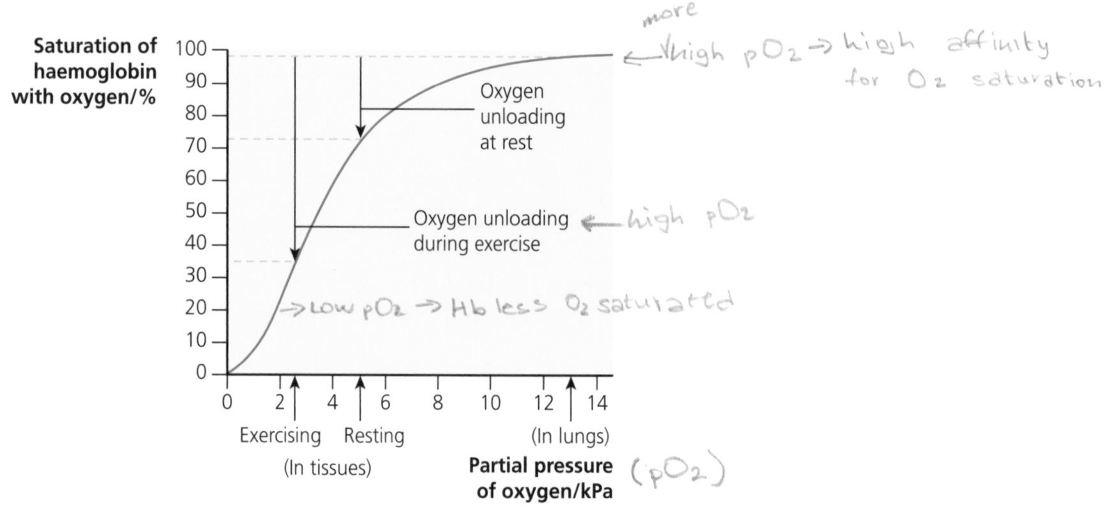

Figure 3.15 The oxyhaemoglobin dissociation curve of human haemoglobin

On the x-axis, the term *partial pressure* is a bit of physics. For our purposes, it means 'how much oxygen is available'. The y-axis is more straightforward. It means 'how many haemoglobin molecules are carrying oxygen?' A value of 80% means that 80% of haemoglobin molecules are carrying oxygen and 20% are not.

So the graph simply shows that haemoglobin will pick up oxygen where it is abundant (the lungs), but it will release oxygen when it is scarce (the respiring tissues in the rest of the body). To do this, haemoglobin must be able to change its affinity for (attraction to) oxygen.

At the lungs, haemoglobin becomes almost fully saturated with oxygen — about 98% is a normal value. Oxygenated blood is then transported to the respiring tissues all around the body. What makes haemoglobin give up its oxygen? When a red blood cell enters a capillary, it begins to unload its oxygen:

1 The respiring cells make carbon dioxide, which diffuses into the blood and into the red cell.
2 The enzyme carbonic anhydrase catalyses the reaction:

$$H_2O + CO_2 \rightleftharpoons H_2CO_3 \rightleftharpoons H^+ + HCO_3^-$$

   This simply means that water combines with carbon dioxide to make carbonic acid. Like all acids, carbonic acid splits to form hydrogen ions and hydrogen carbonate ions. Hydrogen ions are what make solutions acidic.
3 Vitally, the H^+ ions lower the affinity of haemoglobin for oxygen. So haemoglobin releases some of its oxygen molecules.
4 The oxygen molecules are free to diffuse into the cells.
5 The remaining HCO_3^- ions diffuse into the plasma, leaving the red blood cell with a slightly positive charge. To balance this out, chloride ions (Cl^-) diffuse from the plasma into the red blood cell.

Therefore, the higher the **carbon dioxide concentration**, the lower the affinity of haemoglobin for oxygen. It is a really clever mechanism: the faster the cells and tissues are respiring, the faster the oxygen is delivered. If the muscles are working really hard, there will be more carbon dioxide and more H^+ ions, so the affinity of haemoglobin for oxygen is lowered and so more oxygen is released. This is known as **the Bohr effect**. It results in the oxyhaemoglobin dissociation curve moving to the right. The higher the concentration of carbon dioxide, the more the curve shifts to the right, showing that haemoglobin has a lower affinity for oxygen (Figure 3.16).

The Bohr effect is the shift to the right of the position of the oxyhaemoglobin dissociation curve in the presence of extra carbon dioxide.

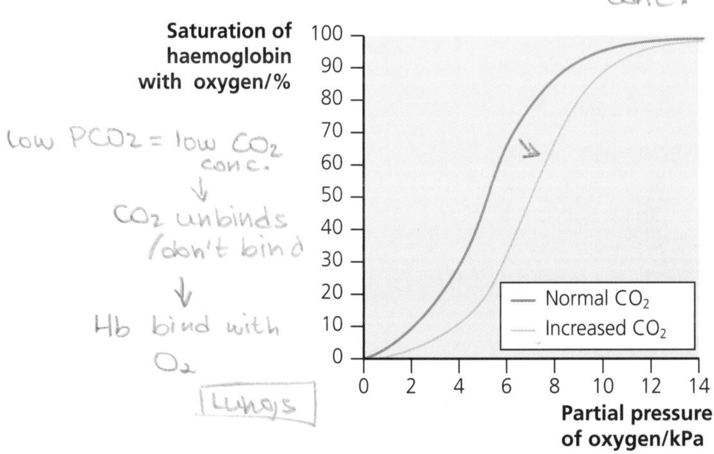

Figure 3.16 The Bohr effect

Exam tip

In questions about haemoglobin, try to use words like *affinity*, *saturation* and *partial pressure*.

Now test yourself

TESTED

16 Explain why red blood cells need the enzyme carbonic anhydrase.
17 Why is it not correct to say that haemoglobin has a high affinity for oxygen?

Answers on p. 204

Haemoglobin in other organisms

There are several examples of different types of haemoglobin in invertebrates. Most do not have a complex circulatory system, nor do they have what we would recognise as blood, but some have haemoglobin in their body tissues. Examples include:

- tubifex worms, which are related to earthworms — they are often called sewer worms and survive in oxygen-poor water
- bloodworms, which are the larvae of a type of midge, often found in ponds and lakes

The key point is that their haemoglobin has a high affinity for oxygen and it can become saturated with oxygen even at low partial pressures. This gives them an advantage: they can respire and survive in conditions of low oxygen that would prove lethal to many other species.

The blood system

REVISED

The **blood circulation** in a **mammal** is closed, meaning that the blood circulates in a complete circuit — it never leaves the **blood vessels**. The sight of blood is a sure sign that a blood vessel has broken. Substances can pass in and out of the blood through capillary walls, but blood stays in blood vessels.

Humans and other mammals have a double circulatory system in which blood flows through the heart twice for each circuit around the body (Figure 3.17):

- the **pulmonary** circulation takes blood to the lungs to take up oxygen
- the **systemic** circulation takes the oxygenated blood around the body to the tissues

body → heart → lungs → heart → body

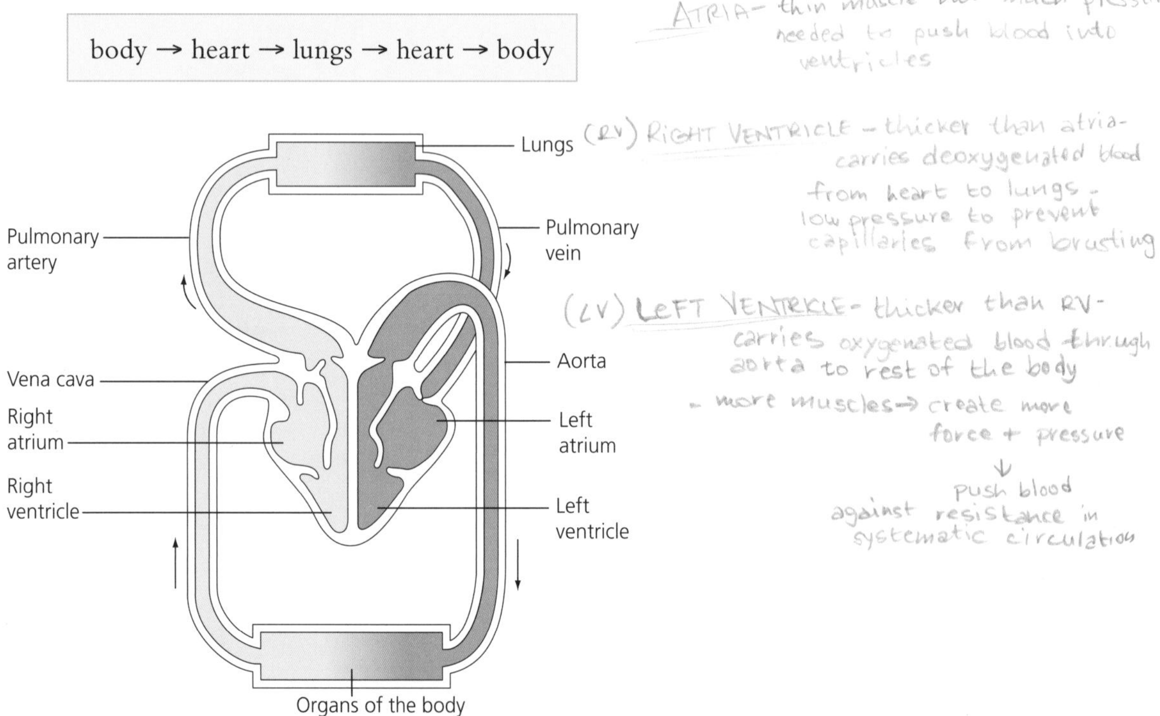

Figure 3.17 The human double circulatory system

We need two circulatory systems because, when the blood goes to the lungs, it takes up oxygen *but loses pressure*. The blood must therefore return to the heart for a boost. It needs to gain enough pressure to allow it to travel around the whole body.

Blood vessels

You need to know about four types of blood vessels: arteries, arterioles, veins and capillaries.

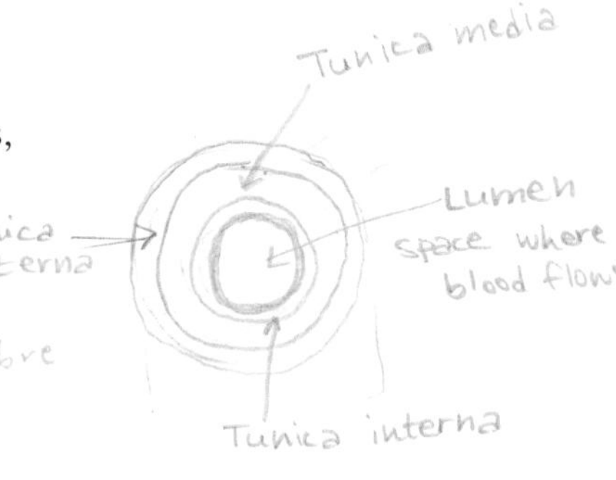

The walls of blood vessels have three layers (Figure 3.18):

- the outer layer — the tough **tunica externa**
- the middle layer — the **tunica media**
- the inner layer — the **endothelium**

It is the difference in the middle layer that gives blood vessels their different properties. Capillaries, however, do not have a middle layer, just a one-cell thick endothelium.

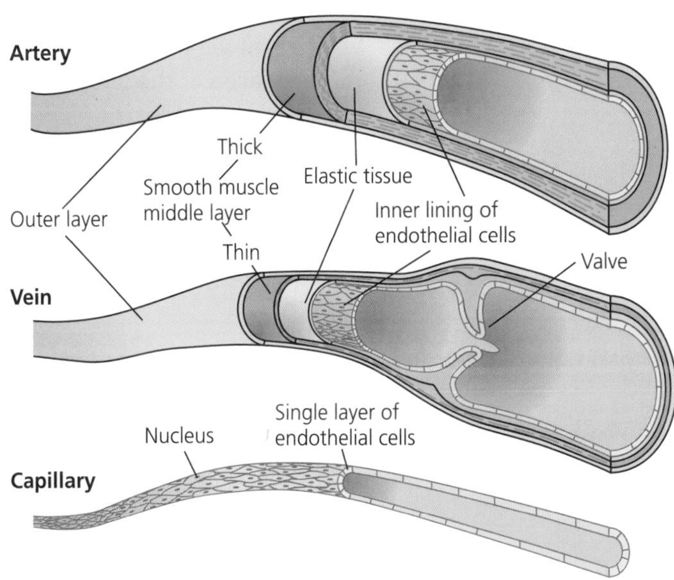

Figure 3.18 The structure of an artery, a vein and a capillary

Arteries

Arteries are adapted to withstand pressure. When the heart beats, the left ventricle forces blood into the body's largest artery, the **aorta**. From here, blood enters the major arteries of the body, leading to all the major organs and limbs. The middle layers of the artery walls are rich in muscle and, vitally, **elastic fibres**. This gives them powerful recoil properties so they can withstand the pressure surge of each heartbeat.

Arterioles

Arterioles are adapted to control blood flow. By the time blood reaches the arterioles, it has lost much of its pressure because it has been absorbed by artery walls. The walls of the arterioles do not need as many elastic fibres, but they do have a lot of circular muscle fibres. This means that arterioles are capable of either:

- vasoconstriction — they get smaller when the circular muscles contract
- vasodilation — they get larger when the circular muscles relax

In this way, blood flow to certain areas of the body can be controlled. For example, vasodilation causes the skin to redden, whereas vasoconstriction causes it to go pale. Viagra makes certain arterioles dilate too.

Exam tip

Muscles *contract*, vessels *constrict*. Don't get them mixed up.

Veins

Veins are adapted to increase blood flow when pressure is low. Compared with arteries, veins have a larger lumen and a thinner wall. This minimises friction so blood can flow more easily. The walls are made of tough connective tissue and there are fewer elastic or muscle fibres. Veins also have valves that can close to prevent backflow.

Capillaries

Capillaries allow exchange between blood and cells. They are the smallest blood vessels and they do not function on their own. Instead, they form a **capillary bed** — an interweaving network of capillaries supplying organs and tissues. Capillary walls (the endothelium) are just one cell thick. The function of capillaries is to allow metabolic exchange of materials between blood and tissue fluid. All living cells in the body are surrounded by tissue fluid.

A **capillary bed** is a network of capillaries that supply blood to a specific organ or area of the body.

Now test yourself

TESTED

18 Name the blood vessels that supply blood to:
(a) the heart muscle
(b) the lungs
(c) the kidneys

19 Explain how the walls of capillaries are adapted to their function.

Answers on p. 204

Tissue fluid

Cells absorb oxygen and nutrients from **tissue fluid**, exchanging them for carbon dioxide, waste products and any substances that the cell makes, such as hormones. It is the function of the circulatory system to keep the composition of tissue fluid as constant as possible. Tissue fluid is basically blood plasma without the cells and large proteins, which are too large to leave the blood.

Two forces are involved in the formation and drainage of tissue fluid (Figure 3.19):

- hydrostatic pressure — the physical pressure of the blood, created by the heart, which forces fluid out of the blood
- water potential — the pressure exerted by dissolved substances in a fluid, which draws water back into the blood

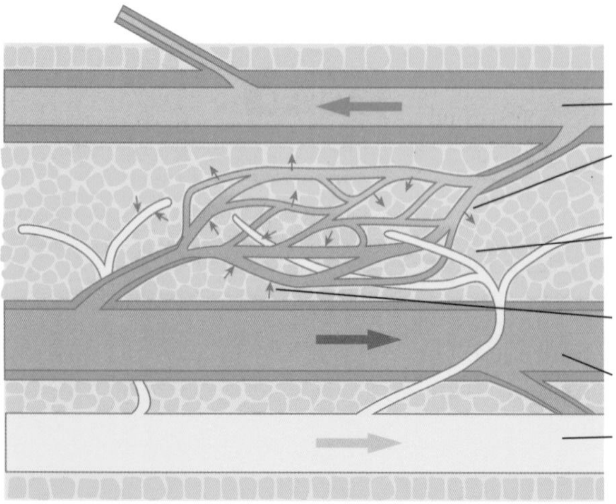

Arteriole (from heart)
Tissue fluid from blood capillaries moves into spaces between cells
Some fluid drains into lymphatic capillaries
Most tissue fluid returns to the capillaries
Venule (to heart)
Lymphatic vessel (to venous system in thorax)

Figure 3.19 The circulation of tissue fluid and the formation of lymph

At the arterial end of a capillary, the hydrostatic pressure is greater than the water potential so that tissue fluid is forced out of the blood. As blood flows along the capillary, it loses volume and therefore loses hydrostatic pressure. However, the proteins that cannot leave the blood exert an osmotic force. So, at the venous end of the capillary, the water potential becomes greater than the hydrostatic pressure and so water is drawn back into the blood (Figure 3.20).

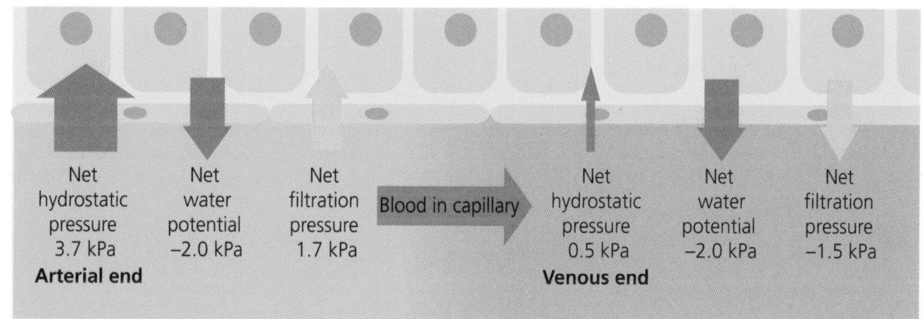

Figure 3.20 As blood flows along a capillary, it loses pressure. When the water potential becomes greater than the hydrostatic pressure, fluid begins to return to the blood

Exam tip

Don't say 'tissue fluid returns to the blood by osmosis'. *Water* returns to the blood by osmosis, while wastes such as urea and CO_2 *diffuse* back into the blood.

Now test yourself

TESTED

20 Name three substances that will pass from blood into tissue fluid.

Answer on p. 204

Lymph vessels

Lymph vessels also drain tissue fluid. The arterioles deliver 100% of the fluid that reaches the tissues, but the veins only take about 99.9% of it back. The remaining tissue fluid passes into lymph vessels that begin in the tissues themselves, so think of lymph vessels as an extra set of veins.

Some tissue fluid, together with some molecules that are too large to pass into the blood, pass into the lymph system. This fluid drains into progressively larger vessels until it drains back into the blood high in the chest cavity, just under the collar bone. This lymphatic drainage is small but vital. If the lymph vessels become blocked — by parasites, for example — the affected area swells up as tissue fluid accumulates. This is known as **elephantiasis**.

Heart structure and function

REVISED

The human heart is a muscular organ with one simple function: to create blood pressure. The heart has four chambers: two **atria** and two **ventricles** (Figure 3.21). The atria are simply there to load the ventricles with the right amount of blood. When full, the ventricles contract powerfully to create the pressure that forces blood into arteries. Contraction of heart muscle is known as **systole** and relaxation is known as **diastole**.

To understand how the heart works, it is important to remember cause and effect. The heart muscle is stimulated to contract, which changes the pressure in the chambers. Blood always flows from areas of high pressure to areas of lower pressure. It is this flow of blood that causes the opening and closing of the valves.

Exam tip

You will not be asked to draw the human heart, but you should be able to label four chambers, four blood vessels and four valves.

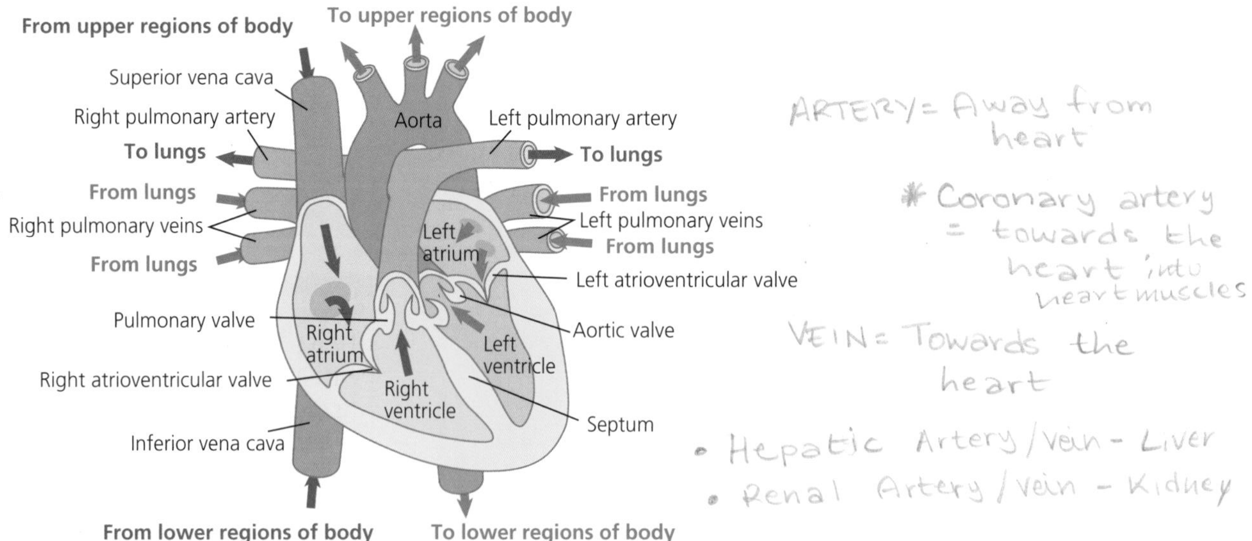

Figure 3.21 The structure of the human heart

Pressure and volume changes during the cardiac cycle

The **cardiac cycle** is the sequence of events involved in one heart beat (Figure 3.22). If your pulse rate is 70, your heart is going through the cardiac cycle 70 times per minute.

The **cardiac cycle** is the series of events in one heart beat.

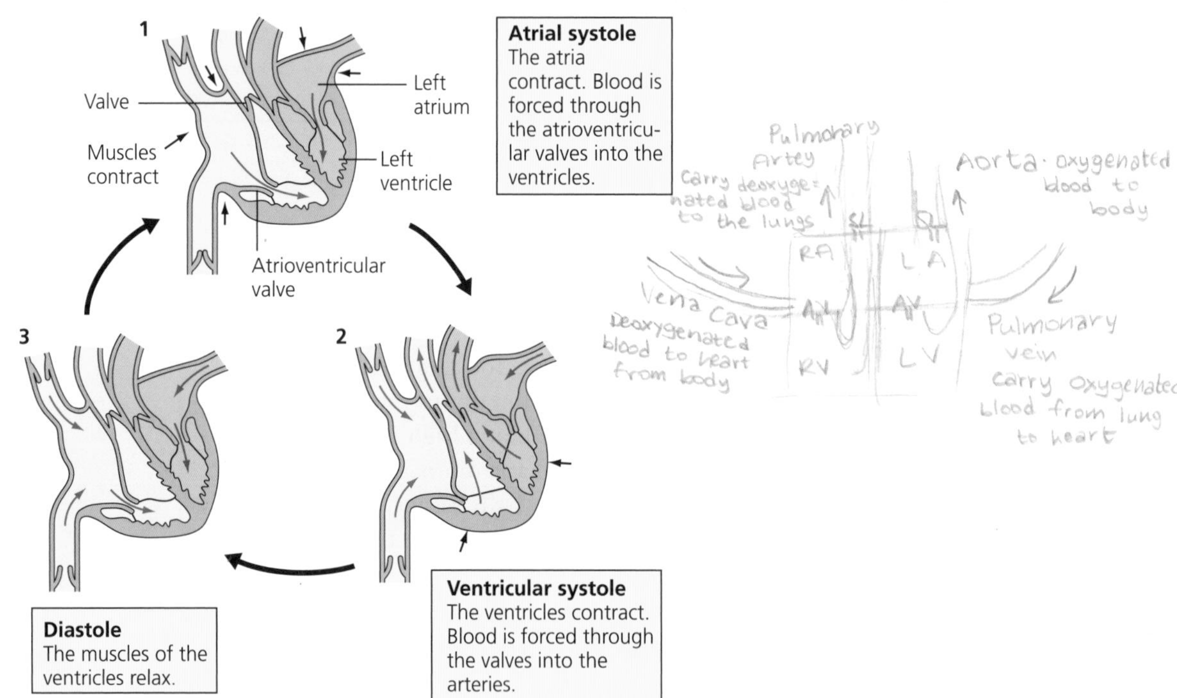

Figure 3.22 The three key stages of the cardiac cycle

The pressure and volume changes and associated valve movements during the cardiac cycle are summarised as follows:

atria fill → atria contract → atrioventricular valves open → ventricles fill → ventricles contract → atrioventricular valves shut → pressure rises dramatically → semilunar valves open → aorta and pulmonary artery fill → as soon as the pressure begins to drop, semilunar valves shut → atria and ventricles relax → heart begin to fill with blood → cycle repeats itself

Exam tip

You should be able to tell the stage of the cardiac cycle by looking at the valves — which are open and which are closed?

Now test yourself

TESTED

21 Put these events of the cardiac cycle in order.
- A Semilunar valves shut
- B Semilunar valves open
- C Atrioventricular valves shut
- D Atrioventricular valves open

22 Is there ever a time in the cardiac cycle when all of the valves are open at the same time? Explain your answer.

23 If the semilunar valves are shut, which chambers are contracting?

24 Which chamber of the heart is responsible for creating blood pressure?

Answers on p. 204

Understanding the cardiac cycle is all about pressure changes and valves. Figure 3.23 shows the pressure changes in one heart beat. There are three lines: the left atrium, the left ventricle and the aorta. Even if there were no labels, you could tell which was which because:

- the pressure in the atria is always low, so it is the bottom line
- the pressure in the aorta (the body's biggest artery) is always high, so it is the top line
- the pressure in the ventricles varies dramatically, so it is the line that rises and falls rapidly

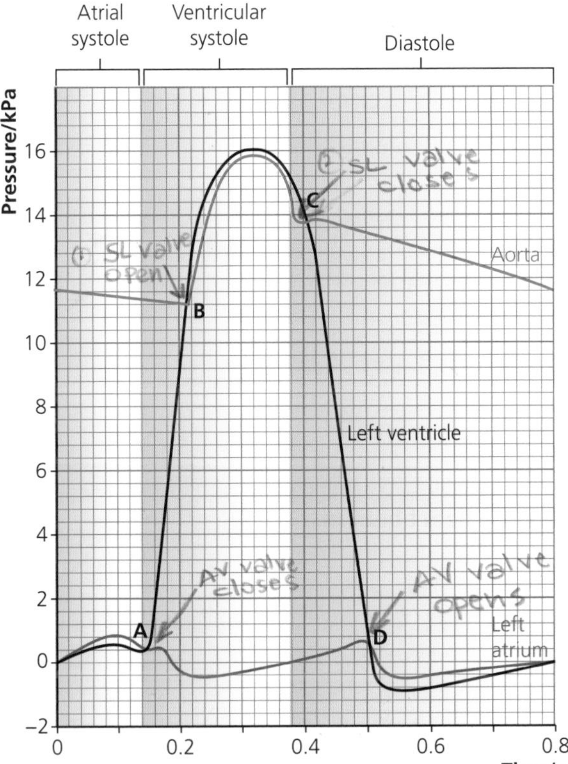

Figure 3.23 Changes in pressure during one cardiac cycle

Now test yourself

TESTED

25 Study Figure 3.23.
- (a) Between which two points is the aorta filling with blood?
- (b) The graph shows the pressure changes in the left side of the heart. Suggest how the graph would be different for the right side.

Answer on p. 204

The conducting pathway of the heart

Heart muscle is myogenic — it is a remarkable tissue that contracts on its own, without nerve impulses from the brain. However, all the cells need to contract at the right time otherwise the heart cannot pump effectively. The chambers should contract only when they are full of blood, so the heart has a conducting pathway of specialised muscle fibres to ensure the right sequence of events (Figure 3.24). The atria must contract first and then, when full, the ventricles follow. This means a delay is needed to allow the ventricles to fill. The full sequence is as follows:

- The heart beat is initiated by a group of cells called the **sinoatrial node (SAN)** near the top of the right atrium.
- These cells produce waves of electrical activity, similar to nerve impulses.
- The impulse spreads over the atria, which then contract.
- A tough band of connective tissue prevents the impulse spreading to the ventricles.
- The impulse is picked up by the **atrioventricular node (AVN)** which, after a short delay, passes the impulse down the middle of the ventricles in the **bundle of His** — a specialised bunch of muscle fibres that transmits the impulse without causing contraction.
- At the apex of the heart (the bottom of the ventricles) the impulses reach the **Purkinje fibres**. These cause contraction of the thick ventricle muscle, starting at the apex so that blood is forced upwards through the semilunar valves.

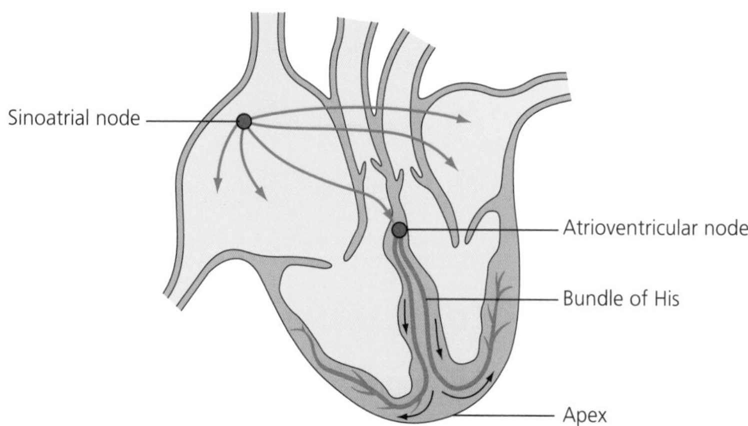

Figure 3.24 The route of electrical activity that makes the heart beat in a smooth sequence

Now test yourself

TESTED

26 Explain what is meant by the term *myogenic*.
27 Explain why it is important that there is a slight delay after the atria contract.

Answers on p. 204

Cardiac output

The volume of blood pumped with each heart beat is called the **stroke volume**. A typical value is around 80 cm^3. This means that each time the heart beats, 80 cm^3 of blood is sent to the lungs via the pulmonary artery and 80 cm^3 is sent to the rest of the body. All of the chambers of the heart have the same volume.

The total amount of blood pumped per minute is known as the **cardiac output**. It is calculated as using this formula:

cardiac output = stroke volume × heart rate

Therefore, the cardiac output for a person at rest might be $80\,\text{cm}^3 \times 70$ beats per minute $= 5600\,\text{cm}^3$, or 5.6 litres.

Now test yourself

TESTED

28 If an athlete has a stroke volume of $100\,\text{cm}^3$ and a heart rate of 160 bpm, work out his cardiac output.

Answer on p. 205

Cardiovascular disease

Cardiovascular disease starts with a build-up of fatty material inside the walls of blood vessels. This may go on for years with no symptoms, but if the arteries get narrower and narrower, a time will come when not enough blood can get through.

Atheroma

Atheroma is fatty material that builds up in the walls of arteries, causing them to get narrower (Figure 3.25). It also causes the artery lining (the endothelium) to get rougher. **Atherosclerosis** is the process of atheroma developing inside the artery walls.

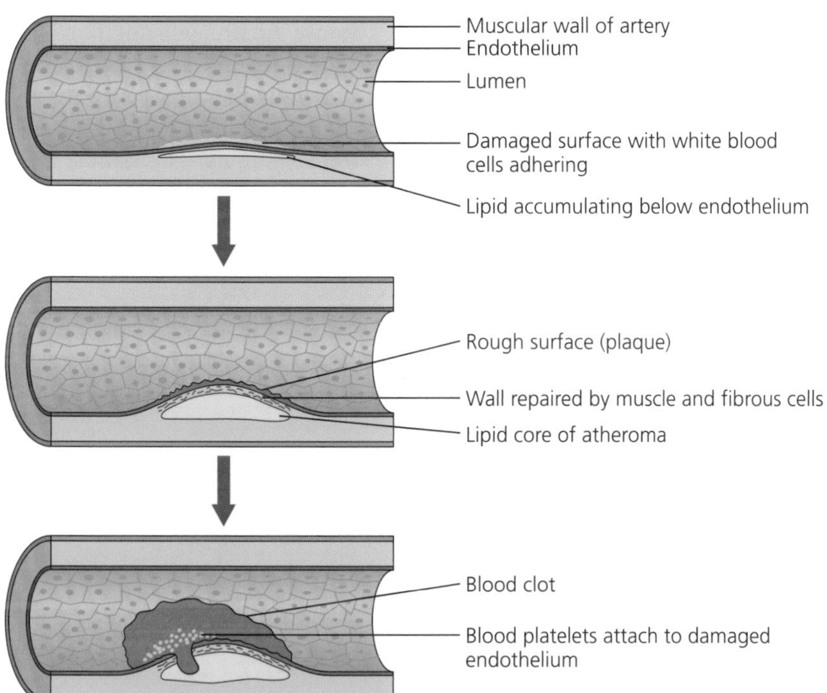

Figure 3.25 The development of atheroma

Typical mistake

Candidates often state that atheroma builds up in the lumen or the artery, implying that the fatty material covers the endothelium. In fact, the fatty deposit accumulates within the artery wall, so that the endothelium is pushed inwards.

Thrombosis and aneurysm

Arteries have to cope with high blood pressure so their walls have to be relatively thick and elastic. The walls of arteries have several layers. The endothelium of a healthy artery should be smooth and unbroken.

Over time, fatty deposits begin to develop under the endothelium. This makes the endothelium rough and sticky, and narrows the lumen. If the

Exam tip

Candidates often get bogged down trying to give the exact components of atheroma, but that level of detail is not required in the specification.

fatty material breaks through the endothelium, it can lead to a blood clot known as a **thrombosis**, which can block the artery. A stroke is caused by a blood clot or burst blood vessel in the brain, so that an area of the brain cells die. The symptoms depend on the size of the damaged area and its precise location in the brain.

An **aneurysm** is a ballooning of the artery that results from a weakness in the vessel wall coupled with high blood pressure (Figure 3.26). The endothelium and middle layer (tunica media) of the artery split so that just the outer layer (tunica externa) is left intact. This is a serious medical crisis that needs surgery. A burst aneurysm will quickly be fatal due massive internal bleeding.

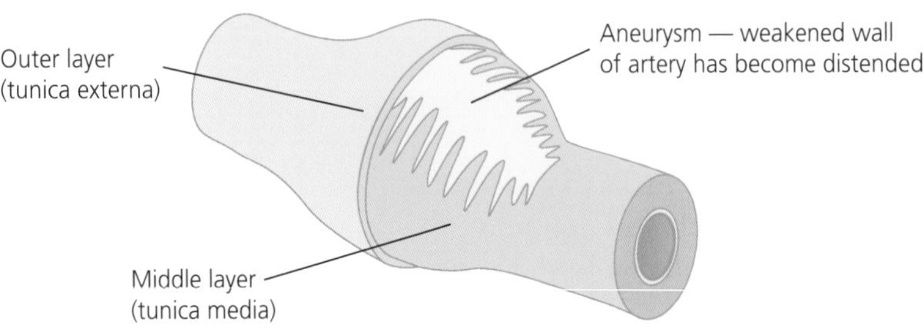

Figure 3.26 An aneurysm

Myocardial infarction

A blocked coronary artery causes a myocardial infarction or heart attack (Figure 3.27). It is caused by an interruption in the blood flow to the heart muscle, so the cells immediately run short of oxygen, cannot respire and therefore die. The symptoms include severe pain in the chest and upper body, particularly on the left side, sweating and shortness of breath. The heart can recover from a small amount of muscle death, but large areas can cause complete heart failure, which is almost always fatal.

Exam tip

Candidates often lose marks by missing out details they know. An answer such as 'The heart muscles cells are starved of blood and so they die' would be much better expressed as 'The heart muscle cells are starved of oxygen, so cannot respire and therefore die'.

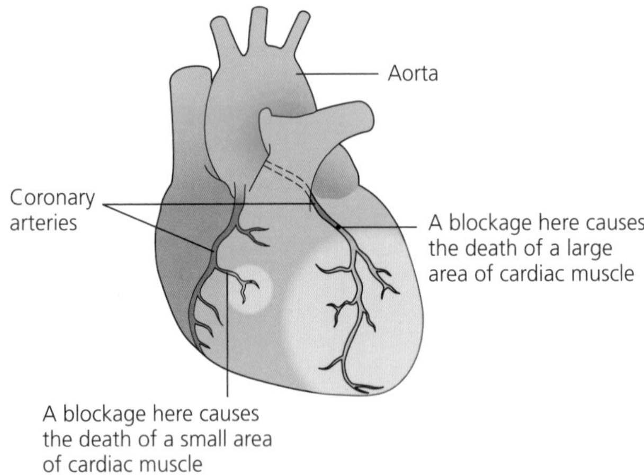

Figure 3.27 A blockage in a coronary artery can cause the death of an area of heart muscle

Risk factors associated with coronary heart disease

Some people are predisposed to develop **coronary heart disease (CHD)**, meaning that they have inherited genes that increase the risk. Lifestyle factors also contribute. These are the main risk factors in the development of CHD.

- Diet — too much saturated fat appears to speed up the development of atheroma. Animal tissue such as eggs, cheese, cream and red meat tends to be high in saturated fats.

- Blood cholesterol — animal fat is also high in cholesterol, which is a major constituent of atheroma.
- Cigarette smoking — there are lots of different substances in cigarette smoke, including nicotine and carbon monoxide. Some of them make the arteries constrict. This reduces blood flow and increases blood pressure. Carbon monoxide binds to haemoglobin and prevents it carrying oxygen. If 20% of your haemoglobin is not working, your heart will have to pump 20% more blood to make up the difference.
- High blood pressure — this is the pressure created in arteries when the heart beats, and it varies according to the strength of the beat and the size and condition of the arteries. High blood pressure can damage artery walls, increasing the risk of atheroma. It also increases the risk of aneurysm and stroke.

Coronary heart disease (CHD) refers to a build-up of atheroma in the coronary arteries that supply the heart muscle with blood.

These are additional risk factors:

- Exercise — having a sedentary lifestyle can increase the chances of developing CHD.
- Weight — a person's weight is usually measured as body mass index (BMI), which is calculated as follows:

$$\text{BMI} = \frac{\text{mass (kg)}}{\text{height}^2\ (\text{m}^2)}$$

  Someone with a BMI of over 25 is classed as overweight, while over 30 they are obese. Obesity brings a whole range of health problems that include CHD as well as type 2 diabetes and joint pain.
- Age — the older we get, the more our lifestyle has a chance to take its toll. Our tissues also get less elastic. Arteries lose their elasticity and are more likely to tear or split as a result of high blood pressure.

In the UK, over 25% of deaths each year are directly due to CHD, making it the biggest killer and pushing cancer into second place.

Now test yourself

TESTED

29 A person is 1.56 m tall and weighs 94 kg. Are they obese? Explain your answer.

Answer on p. 205

Mass transport in plants

Xylem

REVISED

Xylem is a specialised conducting tissue that transports water and dissolved ions up the stem from the roots towards the leaves (Figure 3.28). This example of mass transport is known as the **transpiration stream**. The loss of water vapour from the upper surface of a plant is known as **transpiration**. The vast majority of this water is lost from the stomata, most of which are on the underside of leaves.

The **transpiration stream** is the movement of water and minerals through the plant in the xylem tissue, from the roots to the leaves.

Transpiration is the loss of water vapour from the upper surfaces of a plant.

Xylem vessels develop when cells elongate and then die, so that their contents are lost. Their cell walls become strengthened with extra **cellulose** and a waterproof material called **lignin**. Many xylem vessels join end to end so that a continuous pathway is formed from the root up to the leaves.

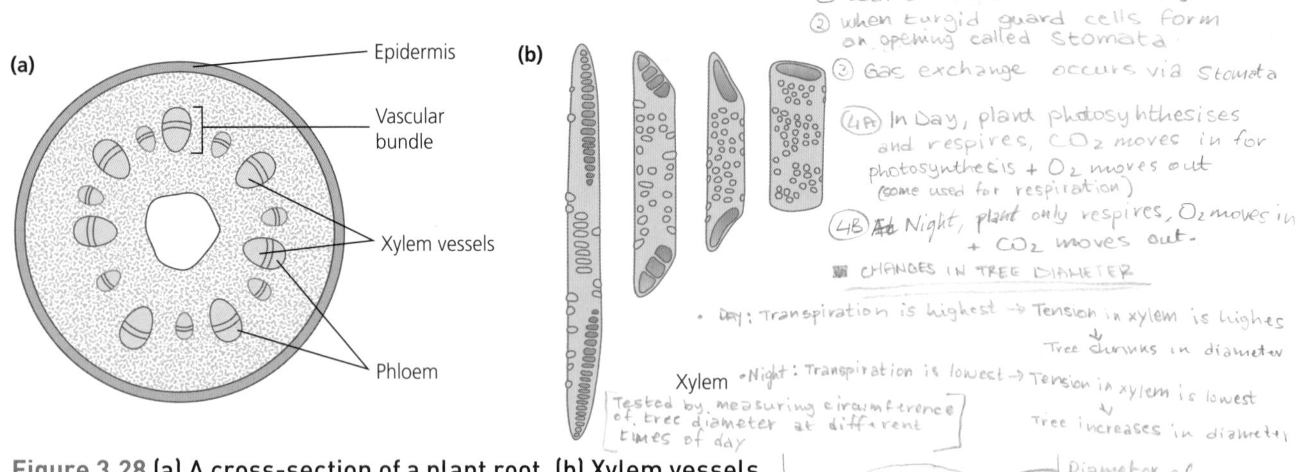

Figure 3.28 (a) A cross-section of a plant root. (b) Xylem vessels

The movement of water up a plant

Leaves must be organs of gas exchange in order for photosynthesis to take place. Carbon dioxide needs to diffuse in and oxygen needs to diffuse out. To allow efficient gas exchange, leaves have stomata and loosely packed mesophyll cells. As a consequence, a large surface area of cells in the leaf is exposed to the atmosphere and there is rapid water loss by evaporation (Figure 3.29). The force that draws water up the plant in the xylem is described below.

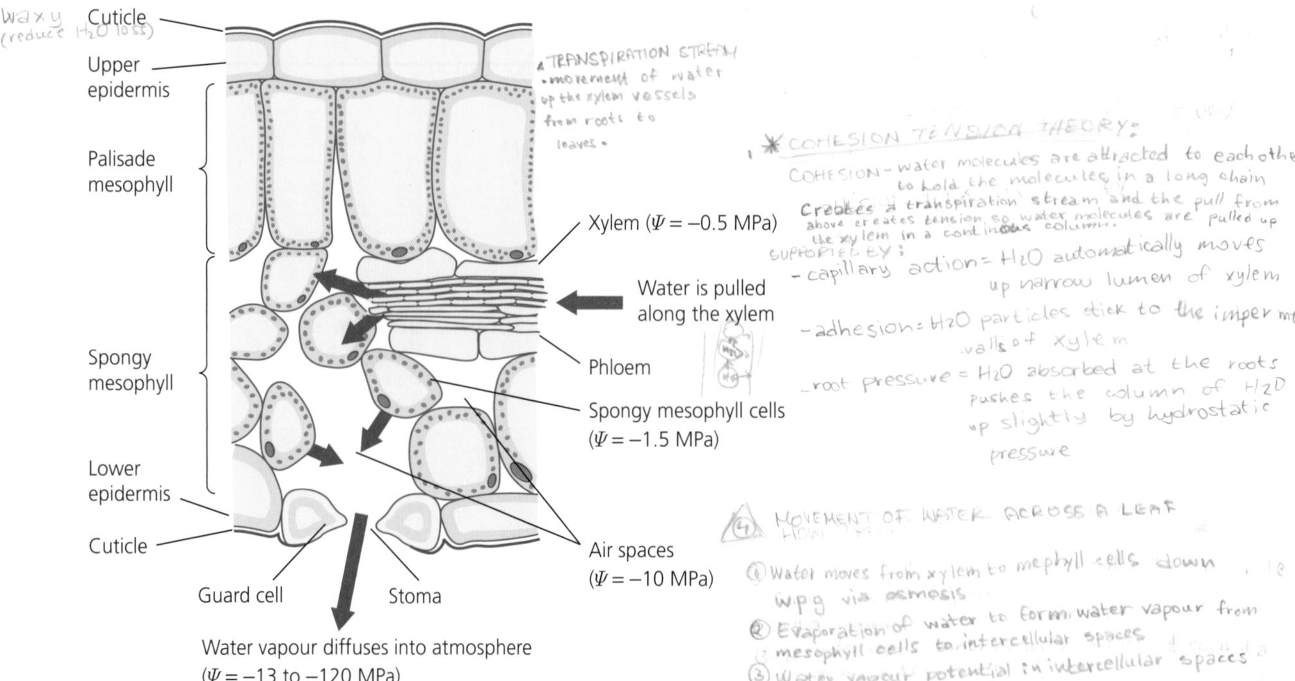

Figure 3.29 The evaporation of water from the leaf creates a pull on the xylem

The cohesion-tension theory

The water potential of dry air is very low. This causes evaporation and is one of the basic forces that drives that water cycle.

1. Water evaporates from the wet cell walls inside leaves and passes out through the stomata.
2. The loss of water creates a pull on the xylem.

3 Water molecules are cohesive (the molecules attract one another). They form a continuous column from the leaves right down to the roots, which can withstand great tension.
4 Therefore, evaporation from the leaves creates a pull that draws water up to the top of the plant.

This is the **cohesion-tension theory**.

Factors that affect the speed of transpiration

Transpiration happens by evaporation, so the conditions that speed up the process are:

- Dry air — the dryer the air, the lower the water potential. This means there is a greater **water potential gradient** between the air inside and outside of the leaf. When air is humid, the water potential gradient is smaller and evaporation is slower.
- Warmth — heat increases the kinetic energy of the water molecules, so they evaporate more quickly.
- Wind — in still conditions, pockets of humid air develop around the stomata. These are called **diffusion shells**. Wind blows away the diffusion shells. This increases the water potential gradient between air and leaf, so when it is windy evaporation is greater.

Exam tip

In questions about practical procedures, examiners often ask candidates to suggest suitable units. However, they sometimes lose marks by forgetting to include 'per unit time'.

Measuring transpiration

The **potometer** is a simple piece of apparatus that measures transpiration — think of it as a transparent extension of the xylem (Figure 3.30). The plant transpires, which draws water up the xylem. In turn, this causes the bubble to move towards the plant, showing how much water is being lost.

The rate of transpiration is measured as the volume of water per unit leaf area per unit time — for example, $3.2\,cm^3\,cm^{-2}\,min^{-1}$. Without leaf area and time, you cannot have meaningful comparisons between different plants, species or conditions.

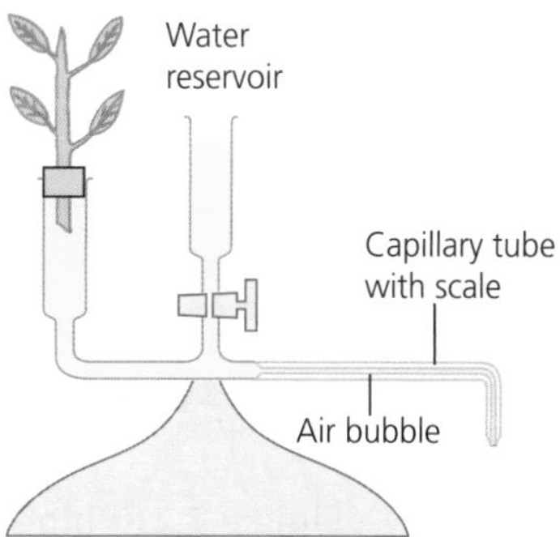

Figure 3.30 A potometer

Now test yourself

TESTED

30 What is the difference between transpiration and the transpiration stream?

Answer on p. 205

Phloem and translocation

REVISED

Translocation is the movement of sugars (mainly sucrose) and other organic molecules around the plant in the **phloem**. There are two important words here:

- **Source** — the part of the plan where the sugar molecules originate. This can be either a photosynthesising leaf making new sugars or storage organs releasing sugars that were made earlier.
- **Sink** — the part of the plant where the sugar molecules are going to. Often this is the roots, leaves, fruits or growing points (meristems — sites of mitosis).

Translocation is the movement of organic molecules (mainly sucrose) around the plant from source to sink in the phloem.

Like xylem, phloem vessels are elongated cells that join end to end to form long continuous tubes (Figure 3.31). There are two basic types of phloem cells: sieve tubes, which actually transport the fluids, and companion cells, which control the activities of the sieve tubes and keep them alive. The differences between xylem and phloem are:

- Phloem vessels are not dead — they contain living cytoplasm. The sieve tubes have a greatly reduced number of organelles and these are concentrated close to the cell wall so as not to impede the flow. The companion cells have a full set of organelles, including nuclei. Xylem vessels have no organelles at all, just walls.
- Phloem vessels move substances in all directions around a plant, whereas xylem vessels just move substances up towards the leaves.
- Phloem vessels contain dissolved organic substances, the majority of which are sugars made by photosynthesis. Xylem vessels just transport water and dissolved ions.
- Movement in phloem is caused by positive hydrostatic pressure, whereas movement in xylem is driven by negative pressure. If you pierced a phloem vessel with a pin, fluid would be forced out. If you pierced a xylem vessel, air would be drawn in.

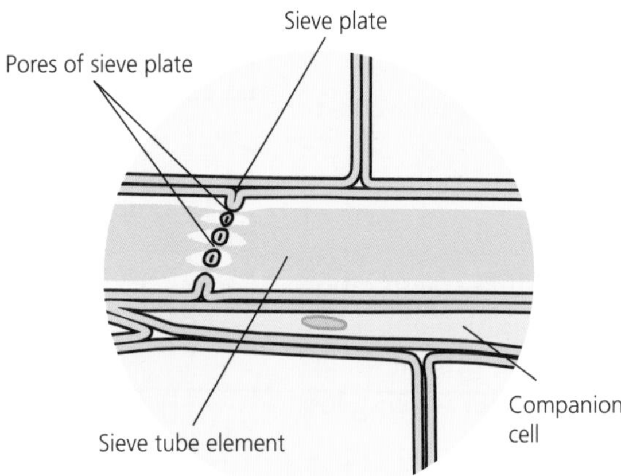

Figure 3.31 Phloem tissue

Now test yourself TESTED

31 What is the difference between translocation and transpiration?

Answer on p. 205

The evidence for translocation

We know that the sugars found in various parts of the plant have originated in the sources and been transported in the phloem. The evidence for this includes **tracers** and **ringing experiments**.

Tracing radioactive ^{14}C

If you provide a plant with carbon dioxide in which the carbon atom is the radioactive isotope ^{14}C, you can follow its progress around the plant using a process called autoradiography. Initially, the ^{14}C is found in the leaves, first in glucose and then starch. Then it can be traced to the phloem, where it is often in sucrose — the most common transport sugar in plants. Finally, the radioactivity can be found in the sinks, often as starch. This is very strong evidence that the phloem is responsible for transporting the products of photosynthesis around the plant.

Ringing experiments

In most trees, the phloem vessels are found on the inside of the bark, whereas functioning xylem vessels are found in the outer part of the wood, just under the bark. If you remove a ring of bark around the circumference of a tree, you remove the phloem. When this is done, sugars accumulate in the phloem above the ring, suggesting that phloem is moving sugars down the plant from the leaves to the roots.

The mechanism of translocation

The **mass flow hypothesis** is the current favoured explanation for the mechanism of translocation:

1 In the sources, sugars are actively transported into the phloem.
2 This lowers the water potential in the phloem.
3 So water enters the phloem by osmosis, increasing the hydrostatic pressure.
4 At the sinks, sugar is actively transported out of the phloem.
5 This raises the water potential in the phloem, so water follows the solute, into the cells of the sink, by osmosis.
6 This lowers the hydrostatic pressure in the phloem in that region.
7 Overall, fluid is forced from the areas of high hydrostatic pressure to the areas of low hydrostatic pressure.

Typical mistake

To avoid confusing xylem and phloem, remember your Fs: *food flows in phloem.* Don't write that in the exam.

Exam practice

1 The graphs show a spirometer trace of the pressure and volume changes in the lungs during one breathing cycle.

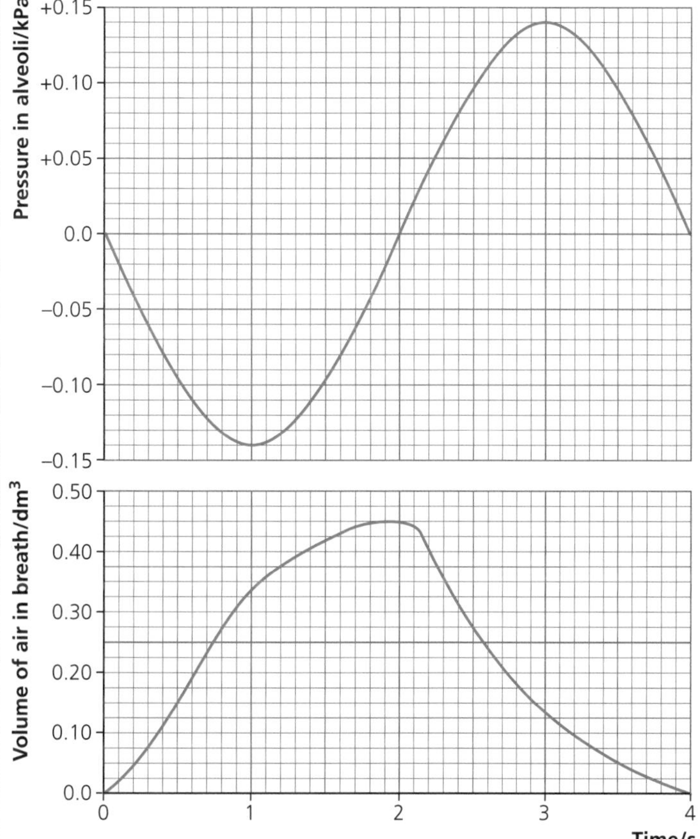

(a) During which time is the person:
(i) inhaling? [1]
(ii) exhaling? [1]

(b) Use the scale to work out pulmonary ventilation rate. [1]
(c) What is value of the tidal volume? [1]
(d) Work out the pulmonary ventilation in litres. [1]

2 The diagram shows three model cells. They are cubes of gelatine with dimensions of 1 mm, 2 mm and 3 mm long.

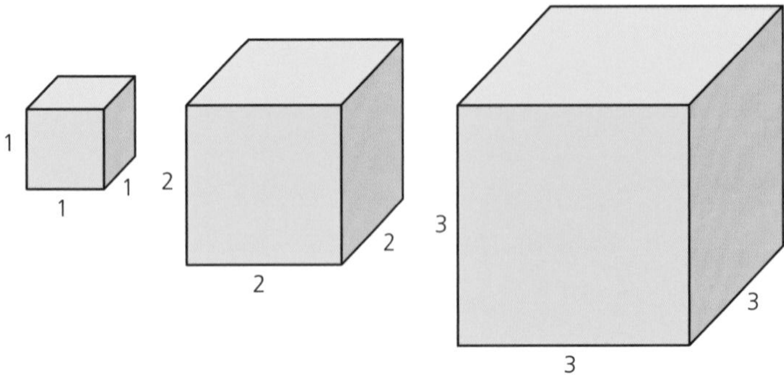

Each model cell was made with gelatine containing a purple indicator that changes to orange in acid conditions. The cubes were placed in weak acid and the time taken for each whole cube to change colour was measured.

(a) The smallest cube weighs 0.002 g. Approximately how much would the largest cube weigh? [1]
(b) Calculate the surface area to volume ratio for the smallest and the largest cubes. [2]
(c) Predict which cube takes the shortest time to change colour when placed in the acid. Explain your answer. [2]
(d) Explain how this investigation illustrates one reason why cells cannot be large. [2]
(e) Evaluate the use of gelatine cubes as model cells. [3]

3 Triglycerides are a major class of lipid. The diagram shows the formation of a triglyceride molecule.

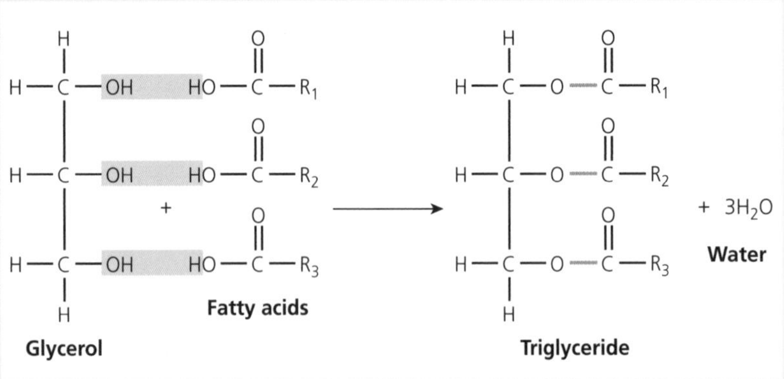

Glycerol + Fatty acids → Triglyceride + $3H_2O$ Water

(a) Give two ways in which the R group of fatty acids can vary. [2]
(b) Name the type of chemical reaction that joins the fatty acids to the glycerol. [1]
(c) Name the bonds that form between the glycerol and the fatty acids. [1]
(d) In total, how many molecules will be formed when the glycerol combines with the fatty acids? Explain your answer. [2]
(e) Describe the difference between a triglyceride and a phospholipid molecule. [2]

4 (a) Explain what is meant by the term *primary structure of a protein*. [2]
(b) The diagram shows cysteine, an amino acid.

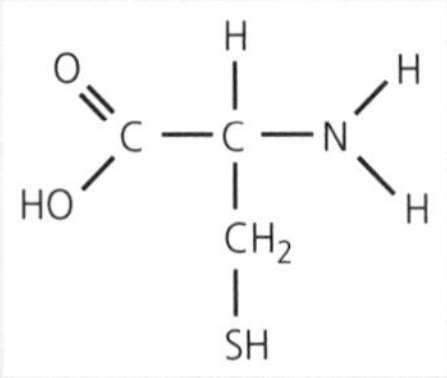

(i) Draw a box around the R group. [1]
(ii) Draw a circle around the amino group. [1]
(c) Disulfide bridges are strong covalent bonds that contribute to structure of a protein. Explain how the amino acid shown can be involved in the maintenance of the tertiary structure. [3]

5 The diagram shows an *E. coli* bacterium, drawn from a micrograph.

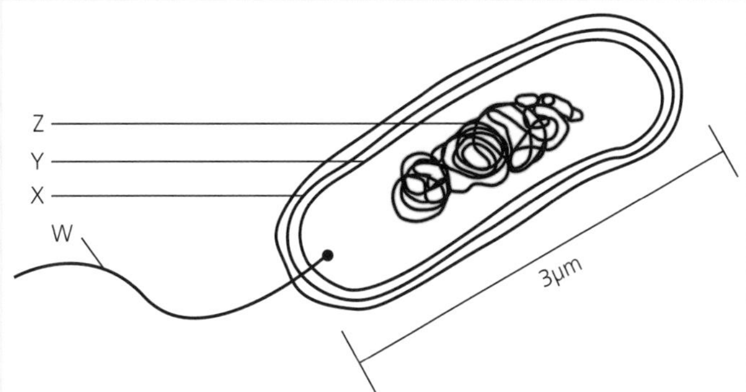

(a) Name the four labelled structures shown. [2]
(b) Name two structures that are normally present in bacterial cytoplasm but are not shown in this diagram. [2]
(c) Suggest why the structures you named in part (b) were not present on the original micrograph. [1]
(d) Use the scale bar to calculate the magnification of the diagram. [2]
(e) This bacterium can swim at a rate of one hundred times its body length per second. Assuming it swims in a straight line, calculate how long it would take the bacterium to cover 1 mm. [2]

6 The table shows the number of deaths due to coronary heart disease per 100 000 population in the UK.

Year	Age 33–44		Age 45–54		Age 55–64		Age 65–74	
	Male	**Female**	**Male**	**Female**	**Male**	**Female**	**Male**	**Female**
1970	65	11	267	46	727	204	1631	704
1980	56	9	270	50	733	215	1621	688
1990	37	6	159	33	536	179	1352	594
2000	19	5	92	20	291	84	823	347
2008	17	4	67	14	175	47	443	179

Source: British Heart Foundation, Coronary Heart Disease Statistics, 2010

(a) Explain why the figures are given per 100 000 population. [2]
(b) Identify three trends shown by the data. [3]
(c) Identify one risk factor for coronary heart disease and explain how it could account for the change in the data. [2]

Answers and quick quiz 3 online

ONLINE

Summary

By the end of this chapter you should be able to understand:

- How the surface area to volume ratio changes as an organism gets larger.
- The essential features of the gas exchange surfaces of single-celled animals, insects, fish and dicotyledonous plants.
- The basic structure of the human gas exchange system.
- The features of the alveolar epithelium as a surface over which gas exchange occurs.
- How the lungs are adapted for gas exchange.
- Pulmonary ventilation rate as the product of tidal volume and breathing rate.
- The mechanism of breathing.
- The effects of emphysema and asthma on lung function.
- The basic structure of the human heart, including the valves and blood vessels.
- The events of the cardiac cycle, the pressure and volume changes and the associated opening and closing of the valves.
- The electrical events of the cardiac cycle, including the roles of the sinoatrial node (SAN), atrioventricular node (AVN) and bundle of His.
- Cardiac output as the product of stroke volume and heart rate.
- The basics of coronary heart disease, including atheroma as fatty material within the walls of arteries.
- The link between atheroma and the increased risk of aneurysm and thrombosis.
- Myocardial infarction and its cause in terms of an interruption to the blood flow to heart muscle.
- The risk factors associated with coronary heart disease.
- The structure of xylem tissue and the idea of transpiration.
- The cohesion-tension theory as the mechanism that draws water up the xylem.
- The factors affecting transpiration and the use of a potometer.
- The structure of phloem and the mass flow hypothesis for translocation.
- Evidence for the mass flow hypothesis.

4 Genetic information, variation and relationships between organisms

DNA, genes and chromosomes

DNA in prokaryotes and eukaryotes

REVISED

This section builds on the structure of DNA and RNA covered on pp. 21–23.

The DNA molecules in prokaryotes are short, circular and not associated with organising proteins. In eukaryotes, the DNA is linear and wound around organising proteins called **histones**. This organisation of DNA and its associated proteins forms **chromosomes**, found in all eukaryotic cells (Figure 4.1).

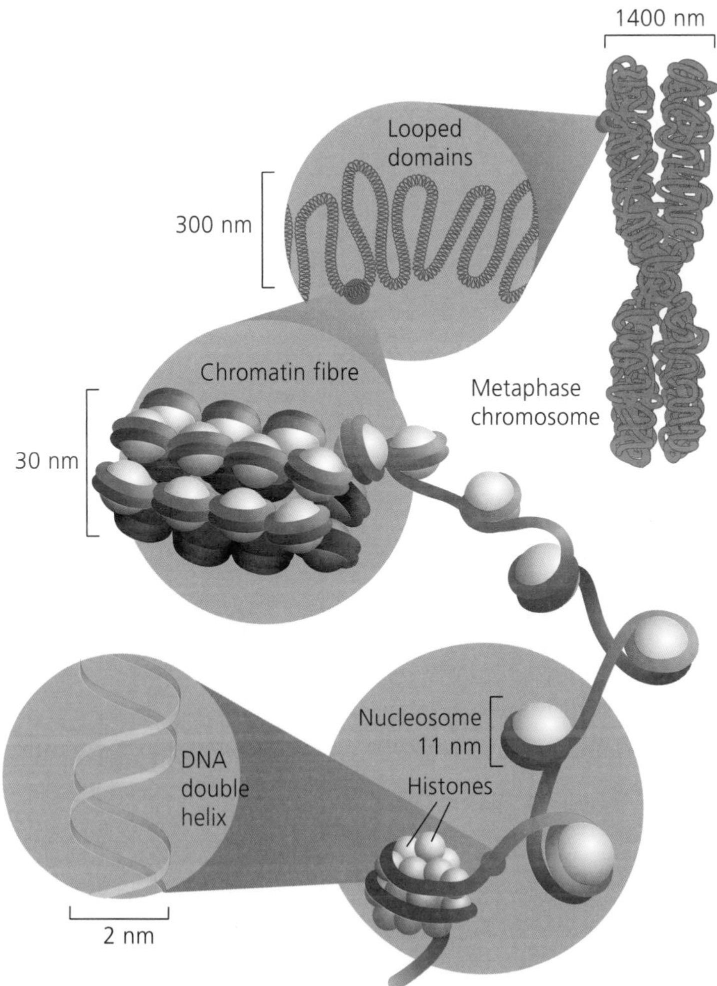

Figure 4.1 After replication, two identical DNA molecules become tightly coiled into a chromosome. Chromatin consists of DNA tightly coiled around organising proteins called histones

Overall, the DNA of **eukaryotes** is found mainly in the nucleus, but there is some in the **mitochondria** and, in plant cells, the **chloroplasts**. The DNA of mitochondria and chloroplasts is circular, which is evidence that these organelles used to be free-living bacteria.

If you look at the nucleus of a non-dividing cell, even with a good microscope, all you can see is, well, mush. In reality, it is a mass of DNA molecules all spread out, so that the genes can be expressed and the DNA can be copied. It is a bit like having the Sunday papers spread out all over the floor. DNA in this state is called **chromatin**. However, when a cell is going to divide it must first copy its DNA and then it must coil it up so it can be equally divided between the new cells. Just like rolling up the newspapers so you can move them around easily.

Each DNA molecule is several centimetres long — an outrageous size for a single molecule. To condense it so that 46 molecules can still fit inside a nucleus requires a lot of organisation — coils within coils within coils. That is what a chromosome is — one long, supercoiled DNA molecule containing hundreds or thousands of genes. The DNA has copied itself exactly, so the chromosomes appear as double structures with two identical sides. Each side, called a **chromatid**, is an exact copy of the other side, joined by a **centromere**.

The total amount of genetic material in an organism is called the **genome**, which includes all the genes and the non-coding sequences in between. Amazingly, the entire genome is present in every body cell. The human genome consists of 23 chromosomes, which occur in pairs. In total, there are over 3 billion base pairs. Different species have different genomes, but it would be wrong to think that the larger the genome, the more complex the organism.

Revision activity

This topic contains a lot of confusing C words. Make a glossary of these words: chromosome, chromatin, centromere, centriole and chromatid.

Now test yourself

TESTED

1. Describe the distribution of DNA in a prokaryotic cell.
2. Describe the distribution of DNA in a eukaryotic cell.

Answers on p. 205

Genes

REVISED

A **gene** is a length of DNA that codes for making a **polypeptide** or a functional **RNA** molecule. A gene is always found at the same position — called a **locus** (plural: loci) — on a chromosome. Different genes have different **base sequences**. A sequence of three DNA bases is known as a **triplet**.

A **gene** is a length of DNA that codes for making one polypeptide or protein.
A gene always codes for making a polypeptide, but some proteins consist of more than one polypeptide, in which case it will be coded for by more than one gene.

Most genes make proteins. The rule is: one gene makes one polypeptide. Sometimes this polypeptide turns into a functional protein and sometimes it needs to be combined with other polypeptides. Haemoglobin, for example, contains two different polypeptides so it is coded for by two different genes. A relatively small number of genes code for essential RNA molecules such as ribosomal RNA (rRNA) and transfer RNA (tRNA).

Proteins produced by genes are essential for growth, repair and the processes of life. Many of them are enzymes or membrane proteins. Our current best estimate is that the human genome contains about 21 000 genes, although there is much we do not know. We need to know where all the genes are, what they code for, how they combine and, crucially, how they are switched on or off in certain cells at certain times.

Non-coding DNA

REVISED

In eukaryotes, most of the DNA does not code for polypeptides. If we could stretch out a single eukaryotic DNA molecule and highlight the genes, we would see that they make up less than 10% of the molecule. There are two areas of non-coding DNA:

- Within genes — non-coding sequences within genes are called **introns** whereas coding sequences are called **exons** because they are 'expressed' to make proteins. Introns must be removed by the cells before the protein is made.
- Between genes — there is a lot of non-coding DNA between genes, often consisting of multiple repeats in which the same base sequence is repeated many times. This DNA varies greatly between individuals and its analysis is the basis of DNA profiling.

Messenger RNA (mRNA) is a nucleic acid that acts as a messenger. It takes a copy of the genetic code from the nucleus into the cytoplasm during protein synthesis. It consists of a single polynucleotide chain with a backbone made of alternating ribose sugars and phosphate groups.

During transcription, the entire base sequence of a gene is transcribed to produce **pre-mRNA** (Figure 4.2). This includes both the introns and the exons. Before it leaves the nucleus, the pre-mRNA is edited and the introns are removed. The exons are then spliced together to produce **mRNA** that carries only the coding sections of the gene.

Typical mistake

Candidates sometimes refer to *extrons* when they mean to say *exons*.

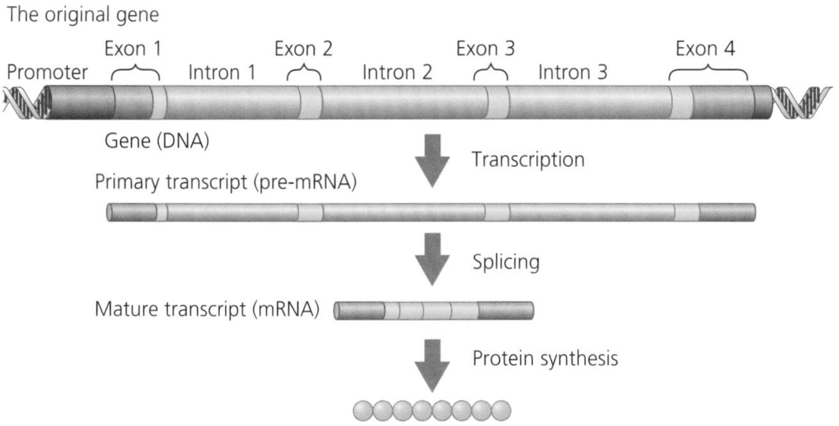

The original gene
Promoter
Exon 1
Intron 1
Exon 2
Intron 2
Exon 3
Intron 3
Exon 4
Gene (DNA)
Transcription
Primary transcript (pre-mRNA)
Splicing
Mature transcript (mRNA)
Protein synthesis

Figure 4.2 Following transcription, the introns are spliced out of the mRNA molecule before it can be translated

The genetic code

REVISED

The **genetic code** is:

- **universal** — the same **codons** code for the same amino acids in all known organisms
- **non-overlapping** — a sequence of CCTGGC is just two codons, CCT and GGC. If the code overlapped there would be codons of CCT, CTG, TGG and GGC. Each base is used once only
- **degenerate** — there are 64 different codons but only 20 amino acids, so there are spare codons. Most amino acids have more than one codon and some, such as leucine, have as many as six (Figure 4.3)

A **codon** is a sequence of three bases on an mRNA molecule that codes for an amino acid, although there are three codons that do not code for any amino acid. They act as 'stop' signals and indicate the end of a particular protein.

First position	Second position: T	C	A	G	Third position
T	Phenylalanine Phenylalanine Leucine Leucine	Serine Serine Serine Serine	Tyrosine Tyrosine (stop) (stop)	Cysteine Cysteine (stop) Tryptophan	T C A G
C	Leucine Leucine Leucine Leucine	Proline Proline Proline Proline	Histidine Histidine Glutamine Glutamine	Arginine Arginine Arginine Arginine	T C A G
A	Isoleucine Isoleucine Isoleucine Methionine	Threonine Threonine Threonine Threonine	Asparagine Asparagine Lysine Lysine	Serine Serine Arginine Arginine	T C A G
G	Valine Valine Valine Valine	Alanine Alanine Alanine Alanine	Aspartic acid Aspartic acid Glutamic acid Glutamic acid	Glycine Glycine Glycine Glycine	T C A G

Figure 4.3 The genetic code

Now test yourself

TESTED

3 How can you tell that Figure 4.3 contains DNA codes and not RNA codes?

4 Use Figure 4.3 to complete this table.

DNA sequence	AAT		GTC
mRNA sequence		GUA	
Amino acid			

Answers on p. 205

DNA and protein synthesis

The **genome** is the complete set of genes in a cell and the **proteome** is the full range of proteins that the cell is able to produce.

The term **proteome** is used to describe all of the proteins that an organism can make. There are many more proteins than there are genes, leading to the idea that by splicing out different introns, one particular gene can make many different proteins.

Synthesising a protein

REVISED

To synthesise a protein, you have to join amino acids in the right order (Figure 4.4). This is exactly what the genetic code does. A protein is a complex molecule that is a long, twisted polymer of amino acids. There are 20 different amino acids that can be joined in any order. These amino acids have names such as valine, leucine, serine and lysine. They are commonly given three-letter abbreviations: val, leu, ser and lys.

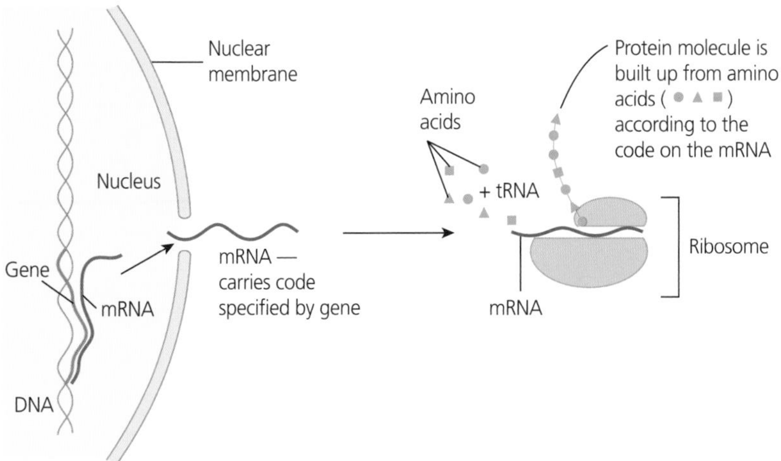

Figure 4.4 An overview of protein synthesis. mRNA is basically a mobile copy of a gene that passes out of the nucleus, taking the genetic code to the ribosome so that the correct protein can be made

Once the amino acids are joined together, many different forces combine to twist and bend the polypeptide chain into a particular shape. The secondary structure refers to structures in part of the molecule, such as helixes and sheets, whereas the tertiary structure is the overall shape of the polypeptide chain.

Now test yourself

TESTED

5 What is the difference between the tertiary and the quaternary structure of a protein?

Answer on p. 205

Complementary bases

REVISED

As there are only four bases but 20 amino acids, the bases are used in sequences of three. A sequence of three bases is known as a triplet, for example CCT or GAA. Each triplet codes for a specific amino acid.

If a protein consists of 100 amino acids, the gene that codes for the protein must have 100 triplets, which is 300 bases. In reality, the gene will be longer because there are usually some non-coding sequences (introns) within the gene as well.

As in DNA replication, the key to protein synthesis is complementary bases (Figure 4.5). When the two strands of DNA are separated, the sequence on the gene can be copied by adding complementary RNA nucleotides. These are similar to DNA nucleotides but the base thymine (T) is replaced by uracil (U). Therefore, if a section of the gene reads TAT GCG TTA, the complementary RNA sequence is AUA CGC AAU.

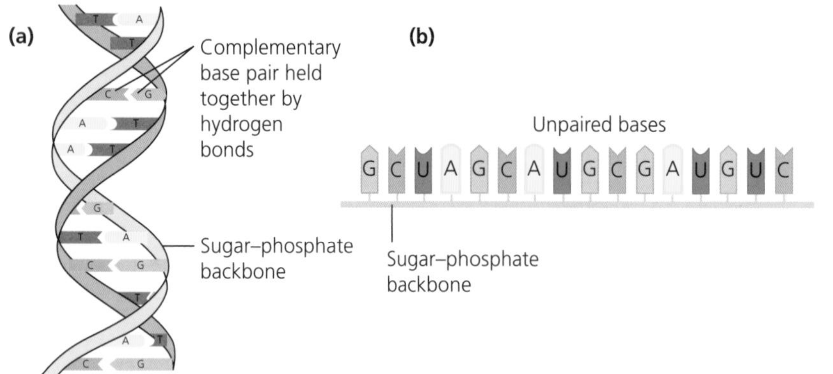

Figure 4.5 The key molecules in protein synthesis. (a) A double-stranded DNA molecule. (b) A single-stranded mRNA molecule

> **Exam tip**
>
> You don't have to remember any codon or amino acid combinations. These will always be provided in the exam question.

The structure of mRNA and tRNA

REVISED

There are two types of RNA in protein synthesis:

- messenger RNA (mRNA) is a long, single-stranded polynucleotide chain that is assembled on a gene
- transfer RNA (tRNA) is a small clover-leaf shaped 'fetch and carry' molecule that brings amino acids to the site of protein synthesis (Figure 4.6)

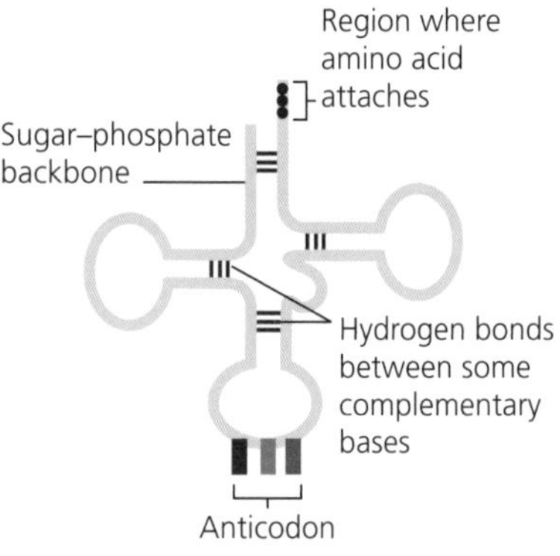

Figure 4.6 The structure of tRNA

Table 4.1 Comparing DNA, mRNA and tRNA

Feature	DNA	mRNA	tRNA
Sugar	Deoxyribose	Ribose	Ribose
Bases	A, C, G, T	A, C, G, U	A, C, G, U
Number of strands	Two	One	One
Hydrogen bonds?	Yes	No	Yes
Number of nucleotides	Millions	Hundreds or thousands — it depends on the size of the gene	About 75

Now test yourself

TESTED

6 Give three differences between the structures of mRNA and tRNA.
7 How many different types of tRNA exist in cells? Explain your answer.

Answers on p. 205

Transcription

REVISED

In eukaryotes

Transcription is the first stage of protein synthesis. DNA cannot leave the nucleus, but proteins are built on the **ribosomes**. As a consequence, the genetic code must be copied in the nucleus and transferred to the ribosomes. Transcription is the process of copying the genetic code by making mRNA from DNA (Figure 4.7).

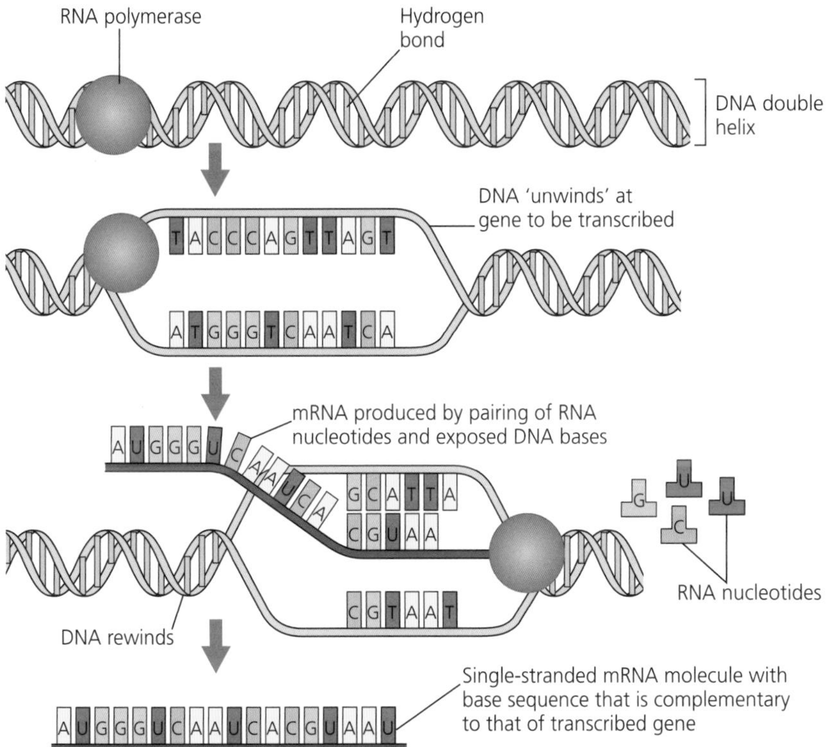

Figure 4.7 Transcription

1. The enzyme **RNA polymerase** attaches to the start of the gene.
2. The DNA unwinds as the hydrogen bonds are broken.
3. The RNA polymerase moves along the gene, catalysing the addition of complementary nucleotides. The entire base sequence is transcribed to produce **pre-mRNA**.
4. Before it leaves the nucleus, the pre-mRNA is edited to form mRNA.
5. The mRNA passes out of the nucleus, through the nuclear pores and into the cytoplasm where it attaches to a ribosome.

Typical mistake

Many candidates state that RNA polymerase adds complementary *bases*, but the correct term is complementary *nucleotides*. It is the bases that actually join, but they also have a sugar and a phosphate attached.

In prokaryotes

In bacteria, transcription is slightly different. The mRNA is made on the gene in a similar process to the one for eukaryotes, but there is no splicing out of the introns. The mRNA passes directly to the ribosomes where it is used to make the polypeptide/protein.

Typical mistake

Many candidates confuse transcription and translation, but make sure you get them the right way round. Remember that *-cription* comes before *-lation* both alphabetically and in biology.

Now test yourself TESTED

8 What is the difference between pre-mRNA and mature mRNA?

Answer on p. 205

Translation

REVISED

In eukaryotes

Translation is the second stage of protein synthesis. It involves assembling a protein by joining amino acids together according to the sequence encoded on the mRNA. The key organelle is the ribosome, which can be thought of as a giant enzyme that holds all the different components together so that the process can happen. The tRNA molecules are relatively small, with two key features:

- an anticodon consisting of three bases
- an amino acid binding site

The anticodon and the amino acid are always matched. For example, the mRNA codon AUG codes for the amino acid methionine. When this codon is translated, a tRNA molecule with the anticodon UAC arrives carrying a methionine at the other end.

1 Ribosomes have two codon-binding sites. The first two codons on the mRNA molecule attach to the binding sites.
2 The first codon is translated. It reads AUG, which codes for the amino acid methionine. A tRNA molecule arrives carrying a methionine. The amino acid is held in place (Figure 4.8).
3 The second codon is translated. The second amino acid is brought in by the tRNA molecule and held alongside the first one.
4 An ATP molecule attaches and is hydrolysed. The energy released is used to form the peptide bond between the two amino acids.
5 The mRNA moves alone the ribosome, one codon at a time. The polypeptide grows as each codon is translated (Figure 4.9).

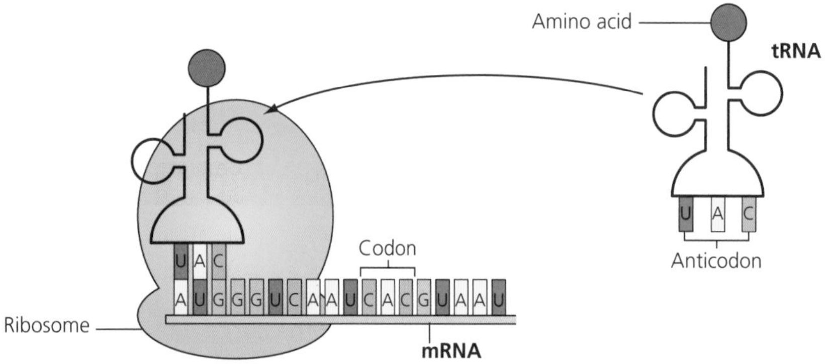

Figure 4.8 Translation, steps 1–2

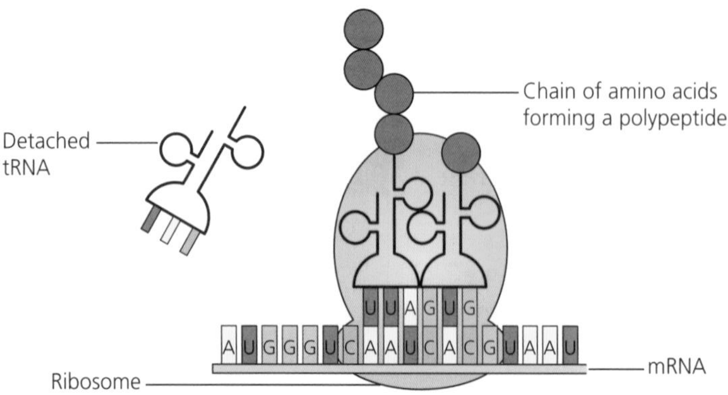

Figure 4.9 Translation, steps 3–5

Revision activity

Draw your own cartoon or storyboard to explain translation — making the components from card works well. You will need a large ribosome, one mRNA with bases, at least two tRNAs with anticodons and an equal number of amino acids.

Now test yourself

TESTED

9 If a polypeptide consists of 62 amino acids, how many nucleotides will the mature mRNA have?
10 Put these events of translation in order.
   A Polypeptide grows
   B mRNA attaches to ribosome
   C Peptide bonds form
   D ATP is split
   E tRNA delivers amino acids and holds them alongside each other
   F The first two codons are translated together
   G mRNA moves along ribosome

Answers on p. 205

Genetic diversity can arise as a result of mutation or during meiosis

A mutation is a change in an organism's genetic material. Mutation can occur at the level of genes or chromosomes.

Gene mutations

REVISED

Gene mutations occur because mistakes in **DNA replication** result in a changed **base sequence**. This changes the genotype of the organism and may be inherited. Mutations do not always affect the organism because:

- some take place in the non-coding DNA between genes
- some take place in the introns (non-coding sequences) within genes
- some will still code for the same amino acid — the genetic code is degenerate. For example, if the codon GUU mutates to GUC, GUA or GUG, it will still code for the amino acid valine
- some will cause a change in the amino acid sequence, but this does not significantly change the tertiary structure. The protein is still the right shape and still functions in the organism

Therefore, the only gene mutations that affect organisms are the ones that bring about significant changes in the structure of the protein. There are two main ways in which a base sequence can be altered:

- **base deletion**, in which one base is lost and there is a frame shift — all the bases move along in one direction and therefore many codons are changed
- **base substitution**, in which one base is substituted for another. This is also called a point mutation. Only one codon is changed, but this can still have a significant effect on the protein

In Figure 4.10, the top two rows show the original sequence. The second two rows show the effect of deleting the red letter A: a frame shift results so that all codons and all amino acids are changed. The third two rows show the effect of a substitution. Here, only one codon and one amino acid are changed.

Original base sequence on mRNA	AGA	UAC	GCA	CAC	AUG	CGC
Encoded sequence of amino acids	Arginine	Tyrosine	Alanine	Histidine	Methionine	Arginine
mRNA base sequence after base substitution	AGU	UAC	GCA	CAC	AUG	CGC
Encoded sequence of amino acids	Serine	Tyrosine	Alanine	Histidine	Methionine	Arginine
mRNA base sequence after base deletion	AGU	ACG	CAC	ACA	UGC	GCx
Encoded sequence of amino acids	Serine	Threonine	Histidine	Threonine	Cysteine	Alanine

Figure 4.10 The effects of mutation

Gene mutations occur randomly. The more times DNA is replicated, the greater the chance that there will be a mutation. Most of these mistakes are spotted and corrected by a 'proofreading' mechanism within the cell. The rate of mutation can be increased by **mutagenic agents**, which include:

- some chemicals including benzene, mustard gas and bromine/bromine compounds
- ionising radiation (gamma and X-rays)
- ultraviolet light
- biological agents such as some viruses and bacteria

Chromosome mutations

REVISED

Chromosome mutations are on a larger scale than gene mutations. If a gene mutation can be likened to changing a word in a sentence, a chromosome mutation can be likened to removing a whole chapter and either throwing it away, inserting it somewhere else, putting it in backwards or attaching it to part of another chapter. As each chromosome contains hundreds or thousands of genes, most chromosome mutations are lethal.

Mutations in the number of chromosomes can arise spontaneously by chromosome **non-disjunction** (meaning 'failure to separate'). During meiosis, each chromosome of a homologous pair should separate and end up in a different **gamete** (Figure 4.11). Non-disjunction results in one gamete getting both chromosomes while the other gets neither.

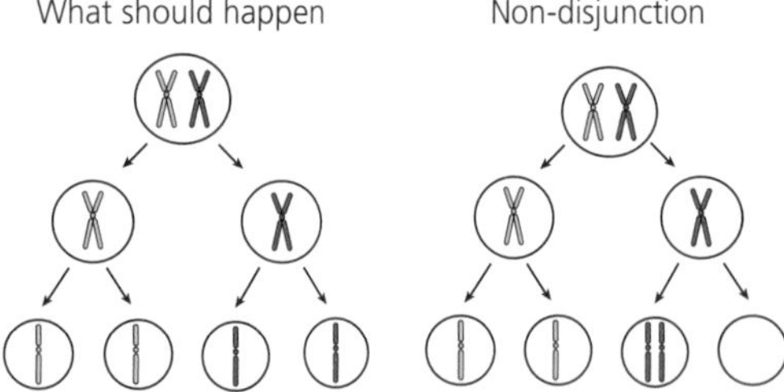

Figure 4.11 Chromosome non-disjunction

Down's syndrome is a common genetic condition caused by non-disjunction of chromosome 21. If it happens in the ovary, one ovum (egg) gets both chromosomes. If this is fertilised by a normal sperm containing one chromosome, the result is three chromosome 21s. If the ovum with no chromosomes 21s is fertilised, the embryo fails to develop. Down's syndrome cannot be described as inherited because it is not passed from parent to child. It is a spontaneous mutation that happens when the gamete is produced. It is a fault of meiosis.

Meiosis

REVISED

There are two types of cell division:

- **mitosis** — one cell divides once to produce two identical daughter cells (see p. 36)
- **meiosis** — one cell divides twice to produce four **daughter cells**, each of which is genetically different and has half the number of chromosomes of the parent cells

Haploid cells contain one set of chromosomes.

Diploid cells contain two sets of chromosomes.

Haploid cells

The only **haploid cells** in humans are the **gametes** (eggs and sperm). They contain 23 chromosomes: one copy of each autosome (22 autosomes) and one sex chromosome (either X or Y). All other cells are **diploid cells**. This means they have two sets of 23 chromosomes each, making a total of 46 chromosomes.

Because our body cells are diploid, this means we have a back-up copy of each gene. If one mutates and does not function, there is another one that does. Diploid cells divide by meiosis to make haploid cells, which fuse to form new, genetically unique individuals.

Typical mistake

Stating that all cells in all organisms have 46 chromosomes — they don't. Human body cells have 46 chromosomes (23 pairs). The cells of most other organisms have a different number of chromosomes.

Homologous chromosomes

There are three golden rules about **homologous chromosomes** (Figure 4.12):

1. They have the same genes.
2. At the same positions (loci).
3. But they may or may not have the same alleles.

Homologous chromosomes are a pair of chromosomes that carry matching genes.

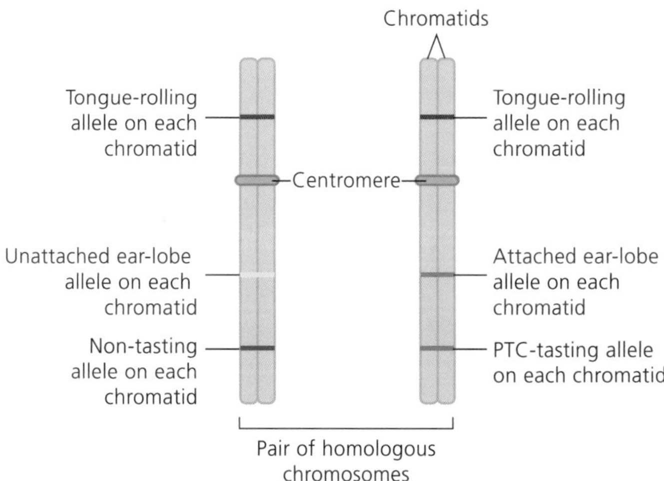

Figure 4.12 A pair of homologous chromosomes with a few genes labelled. Note that the same genes are always at the same loci, but sometimes the alleles are different

Exam tip

Meiosis is a complex process that goes through two divisions. You don't need to know the individual stages, just the ways in which it causes variation. For example, you don't need to know about prophase 1 or metaphase 2.

Genetic recombination

Meiosis creates variation via two processes:

- **Crossing over** (or simply '**crossover**') — homologous chromosomes line up alongside each other and swap blocks of genes. Points of attachment (**chiasmata**) form and sections of chromosome swap between paternal and maternal chromosomes. This process creates new allele combinations.

- **Independent segregation** — at random, one from each of a pair of chromosomes passes into the daughter cell. If there were three pairs of chromosomes, there are 8 (2^3) different combinations of chromosomes that can result.

Once meiosis has created large numbers of unique gametes, a third process guarantees even more variation: **random fertilisation**. Any sperm can fertilise any egg, with the result that every individual is unlike any who has existed before or ever will again.

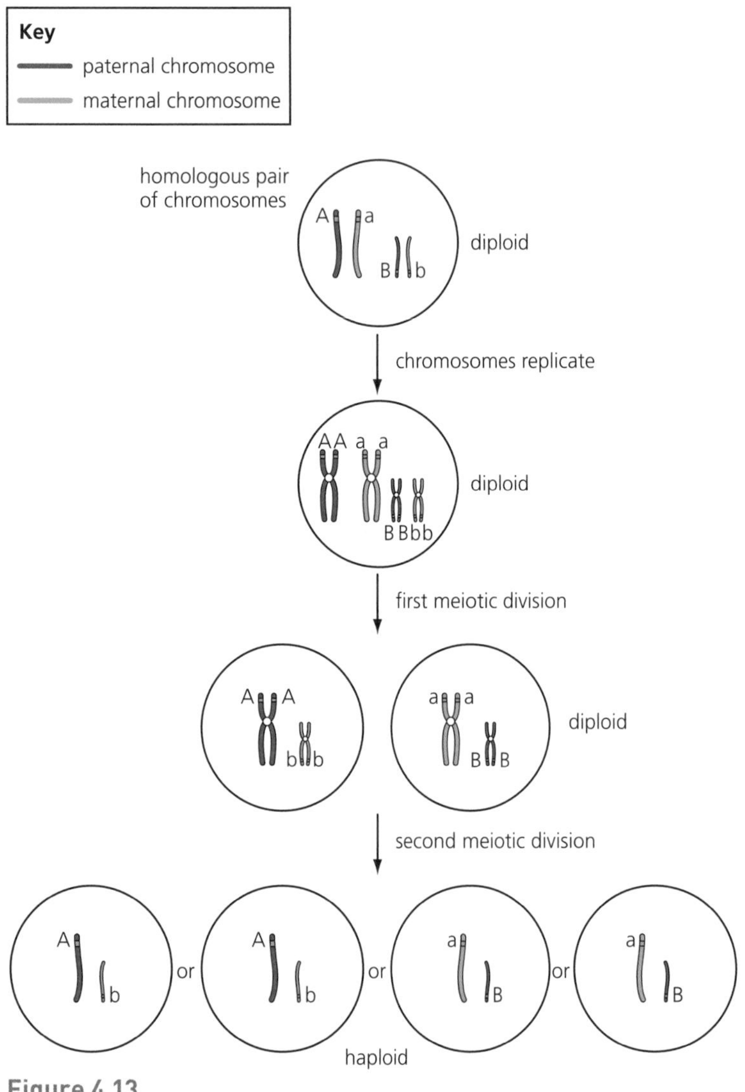

Figure 4.13

Now test yourself

TESTED

11 Human cells have 23 pairs of chromosomes. How many different combinations of chromosome pairs can result from independent segregation in meiosis?

Answer on p. 205

Genetic diversity and adaptation

Genetic diversity can be defined as the number of different **alleles** of **genes** in a population. Without genetic diversity there can be no natural selection and no evolution. If all organisms in a population had the same alleles, they would all be clones, and that is bad.

Natural selection

REVISED

The idea that species can change over time has been around for centuries, but Charles Darwin was first to come up with a workable mechanism for how the change can actually happen. That mechanism is **natural selection** and, arguably, it is the most important idea in biology. The key points are:

- New alleles are created by **random mutation**.
- Many mutations are harmful but, occasionally, an allele or combination of alleles gives certain individuals an advantage and leads to increased **reproductive success**.
- These individuals pass on their advantageous alleles to the next generation.
- Over many generations there will be a change in the frequency of these new alleles in the population. **Evolution** is defined as a change in allele frequency in a population.

Typical mistake

In exam questions, candidates often state or imply that organisms evolve in their own lifetime — they don't. It is all about luck. Those that are born lucky, with the right alleles or combinations of alleles, will pass on more of their alleles to the next generation.

Natural selection results in species that are better adapted to their environment. These adaptations can be:

- **anatomical**, such as a thicker shell
- **physiological**, such as resistance to a particular pesticide
- **behavioural**, such as social cooperation

Typical mistake

Natural selection is often seen as a matter of life or death, and candidates sometimes state that a poorly adapted individual will die and that stops them reproducing. However, it is often more subtle than that. It is about reproductive success. A better adapted individual reproduces more successfully than a less well adapted one, even though they might live in the same population for years.

Two types of selection

REVISED

There is a tendency to think that natural selection always leads to evolution, but it can be a force for stability, giving us two basic types of selection: directional and stabilising.

Directional selection

Directional selection (Figure 4.14) is force for change. In this case, extremes of phenotype have an advantage, such as the thickest fur or the deepest roots. A simple example is antibiotic resistance in bacteria. Individuals that possess the resistance allele will survive and so more of the next generation will possess the resistance allele.

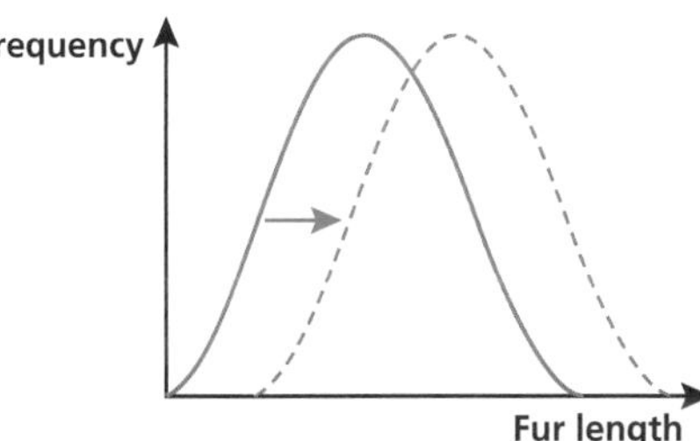

Environmental changes favour the selection of longer fur, causing the normal distribution to shift

Figure 4.14 Directional selection

Stabilising selection

Stabilising selection (Figure 4.15) is seen in cases where the individuals at the extremes of phenotype are at a disadvantage compared with those in the mid-range. Birth weight in humans is a classic example of this. Very large babies cause problems in childbirth whereas very small babies have a lower chance of survival due to a variety of reasons, including a reduced surface area to volume ratio (making them vulnerable to heat loss) and a weakened immune system.

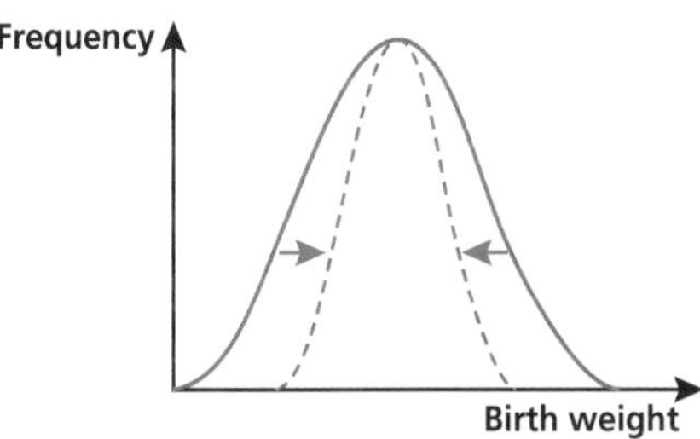

In a stable environment, selection operates to reduce the numbers of heavy and light babies born

Figure 4.15 Stabilising selection

Now test yourself

TESTED

12 Suggest why stabilising selection tends to reduce diversity.

Answer on p. 205

Species and taxonomy

What is a species?

REVISED

A **species** is a group of organisms with observable similarities that can interbreed to produce **fertile offspring**. This definition works well enough for most situations, including A-level exams. It is well known that horses and donkeys can mate to produce sterile hybrids called mules (or asses), and that lions and tigers can produce ligers or tigons, which are also sterile.

However, there are problems with this definition. For example, wolves, coyotes and domestic dogs are all classed as separate species and all have their own scientific names, but they can all interbreed to produce fertile offspring.

Now test yourself

TESTED

13 What is the definition of a species?

Answer on p. 205

Courtship behaviour

REVISED

In most species of animals, **courtship behaviour** is necessary to ensure **successful mating**. Courtship always involves some type of signal, such as:

- visual — displays, dances, flashes of light
- noises/songs
- pheromones — chemical messengers
- tactile — involving touch

There are several reasons for courtship:

- to find a member of the same species (**species recognition**)
- to approach safely and without aggression — male spiders and praying mantises need to get it right to avoid being eaten, for example
- to choose a strong and healthy mate
- to form a pair bond — many species cooperate over parental care and some mate for life

Not only does courtship allow organisms to find each other, it also helps scientists to identify different species. For example, there are several different species of firefly that look exactly the same, but they have different 'flash' patterns. Finding the right species is important because it avoids infertile matings. If the sperm and egg are not from the same species, they will not be chemically compatible and fertilisation will not take place.

Phylogenetic classification

REVISED

The phylogenetic classification system is all about finding out what organisms evolved from what. It is a system of classification based on the evolutionary origins and relationships between organisms and groups of organisms. To construct a phylogenetic tree, scientists use:

- anatomical features, such as body plans
- fossils
- biochemical analysis of base sequences in DNA or amino acid sequences in proteins

Figure 4.16 shows a phylogenetic tree for humans and great apes. For any two species, the golden rule is: the more closely related, the more recently they had a common ancestor. This is the point at which the two species began to evolve along different lines.

Common ancestors are shown by junctions (or nodes) in the diagram. Figure 4.16 shows that the closest relative to humans is the chimpanzee. Gibbons are more distantly related, so humans and gibbons had a common ancestor further back in time.

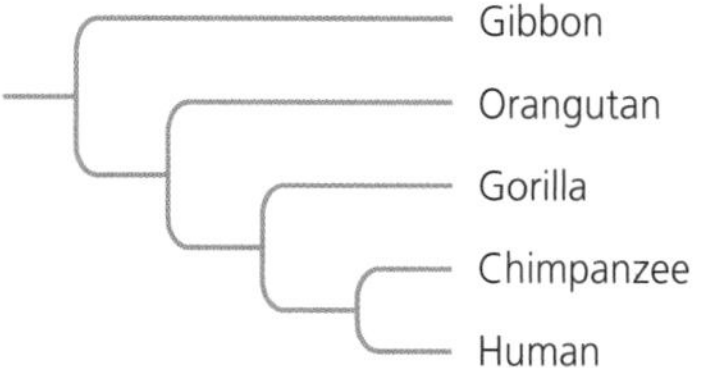

Figure 4.16 The phylogenetic tree for humans and great apes

Now test yourself

TESTED

14 List all the relatives of the gibbon, starting with the closest.

Answer on p. 205

Taxonomy

REVISED

Taxonomy is the science of classification. Scientists are working towards a giant family tree showing the history of life on Earth — what evolved, from what, and when. This will probably never be completed because the extinct species far outnumber the living ones and most of them left no fossils.

Classification is one area of science where there is very little agreement. Systems and ideas are changing constantly as new evidence comes to light. Modern DNA techniques and protein analysis give new evidence about **evolutionary origins** and **relationships**, so that our ideas are changing constantly.

Taxonomic hierarchy

A **hierarchy** is a layered system. The key points of the taxonomic hierarchy are:

- It consists of a series of eight groups, from the most general (domain) to the most specific (species). Similar species are placed in a genus. Similar genera are placed in a family etc. Each group is called a **taxon** (plural: **taxa**).
- The eight groups are **domain**, **kingdom**, **phylum**, **class**, **order**, **family**, **genus** and **species**.
- There is **no overlap** between the groups. For example, there is no organism that is part fish and part amphibian — it is either in one group or the other.
- The groups are based on shared features. The more specific the group, the more shared features there are.

For example, humans and jellyfish are both in the animal kingdom (Table 4.1) because they both share the basic features of animals: they are multicellular, have cells with no walls, move in search of food and digest it in a gut. There, the similarities end.

Revision activity

Think up your own memorable phrase featuring the letters DKPCOFGS to help you remember the groups.

Table 4.1 Classifying humans

Taxon	Humans as an example	Reasons
Domain	Eukaryota	We are organisms whose cells have a distinct nucleus
Kingdom	Animalia	We are animals
Phylum	Chordata	We have a nerve cord — the sub-phylum vertebrata contains animals that have a backbone
Class	Mammal	We have fur and feed our young on milk
Order	Primate	We are social animals with a large brain, opposable thumbs, finger nails, and colour binocular vision
Family	Hominidae	We are man-like apes
Genus	*Homo*	*Homo* means man — there is only one living species in this genus
Species	*sapiens*	'Wise/thinking' man

The binomial system

REVISED

Currently, there are over 2 million different species that have been discovered and given a scientific name. Many more are yet to be discovered.

Each species is universally identified by its scientific name, which is **binomial** — each name has two parts — the lion is *Panthera leo*, for example — which usually come from Latin or Greek. This name is used across all language barriers and avoids confusion when referring to a particular species.

The first part of the scientific name is the genus, which always has a capital letter. The second part is the specific, or species, name, which is always lower case. When writing scientific names, they should be underlined or in italics.

Now test yourself

TESTED

15 The binomial name for the cat is *Felis catus*. Which is the genus and which is the species name?

Answer on p. 205

Biodiversity within a community

Biodiversity is a measure of the richness of a habitat or ecosystem. It can apply to a range of **habitats**, regardless of the size or shape of that habitat.

Index of diversity

REVISED

Biodiversity can be given a numerical value, known as an **index of diversity**, as a result of two measurements:

- the number of species in a **community** (**species richness**)
- the number of individuals of each species (**species evenness**)

Therefore, many individuals from many different species means there is a high diversity. Two classic examples of high-diversity ecosystems are tropical rainforests and coral reefs.

Species richness is the number of different species in a habitat.

Species evenness is how evenly each species is represented throughout a habitat.

Now test yourself

TESTED

16 List two pieces of information needed to calculate an index of diversity.

Answer on p. 205

An index of diversity is a useful measure of the health of an ecosystem. It can also be used to compare one ecosystem with another or to see if anything is changing from year to year. If the index is calculated again at the same time of year and using the same sampling techniques, any difference in value indicates changing conditions. In particular, a lower value would be a worry, suggesting that conditions are deteriorating. Possible causes for change include:

- climate change, such as warmer water causing increased acidity (lower pH)
- overfishing of coral reef fish for food or the aquarium trade
- the introduction of foreign species that out-compete the native species
- pollution

Calculation of an index of diversity

There are several versions of the formula for calculating an index of diversity, so make sure you are consistent. In this case, we will use the formula:

$$d = \frac{N(N-1)}{\Sigma n(n-1)}$$

where:

N = the total number of individuals in all species

n = the total number of individuals in a particular species

Example

Look at the numbers of fish from a coral reef in Table 4.2.

Table 4.2 **Data on fish numbers in a coral reef**

Species	Number of individuals
Clown fish	16
Butterfly fish	9
Four-spot butterfly fish	5
Queen angelfish	2
Koran angelfish	1
Clown triggerfish	2

If an exam question asks you to work out an index of diversity, one approach is to add extra columns to carry out further calculations (Table 4.3).

Table 4.3 **Calculating an index of diversity**

Species	Number of individuals (n)	$n - 1$	$n(n - 1)$
Clown fish	16	15	240
Butterfly fish	9	8	72
Four-spot butterfly fish	5	4	20
Queen angelfish	2	1	2
Koran angelfish	1	0	0
Clown triggerfish	2	1	2
Total	**35**		**$\Sigma = 336$**

Substituting the values into the formula:

$$d = \frac{N(N-1)}{\Sigma n(n-1)}$$

$$= \frac{35 \times 34}{336}$$

$$= \frac{1190}{336}$$

$$= 3.54$$

Therefore, the index of diversity in this example is 3.54.

The effects of farming on biodiversity

REVISED

Farming techniques can reduce biodiversity. In the UK, for example, farming involves the increasing use of technology. Agricultural practices that can affect biodiversity include:

- **Deforestation** — although most deforestation took place in the UK centuries ago, it is a massive worldwide problem. Removal of trees obviously reduces tree diversity as well as removing vital niches for a wide variety of other species.
- **Monoculture** — supermarkets want to buy from just a few large suppliers rather than lots of small ones. It makes economic sense to have large fields growing just one crop, such as wheat or potatoes, which can be harvested efficiently using large machines.
- **Removal of hedgerows** — this turns small fields into large ones, providing more space for crops and making it easier to use large machines. Hedgerows are important habitats for many native species. It used to be a widespread belief that hedgerows also harboured pests and competed with the crops for nutrients, but recent research shows that hedges actually help crops and are just as likely to contain species that eat pests.
- **Use of pesticides** — herbicides kill weeds that compete with the crop for light and nutrients whereas insecticides kill insect pests. It is difficult to make and apply a pesticide that will kill only the pest and nothing else.

Now test yourself

TESTED

17 List four human activities that reduce biodiversity.
18 Give two reasons why deforestation reduces biodiversity.
19 Explain how the eradication of weeds leads to a reduction in animal diversity.

Answers on p. 205

Investigating diversity

Modern technology allows us to study the DNA and proteins in organisms, which is much more precise than studying body plans or fossils. Advanced sequencing techniques allow us to see how closely related two species are by comparing the base sequences in the DNA or the amino acid sequences in proteins. The basic idea is simple: the more sequence differences there are, the more distantly related the species.

Mutations and changes to the DNA base sequence

REVISED

The sequence differences come from the process of mutation. The more distantly related the two species, the more time that has elapsed since they had a common ancestor and the more opportunities there have been for mutations to occur.

Mutations change the DNA base sequence. Sometimes, changes in the base sequence also lead to a change in the amino acid sequence, but not all mutations change amino acid sequences because:

- most DNA occurs in between genes — it is non-coding DNA
- there is some non-coding DNA within genes in sections called introns
- sometimes a changed base sequence still codes for the same amino acid — some amino acids can have as many as six different triplet codes

Vitally, the mutations that do change amino acid sequences must be subtle. The changes must be minor otherwise the tertiary structure would be changed and the protein would not work. A non-functional protein is usually a major disadvantage and so the mutation dies with the organism. Therefore, only these very minor changes to proteins accumulate over time.

In order to compare proteins, you should study one that all organisms have, such as **cytochrome C**, which is found in the inner mitochondrial membrane and has a vital role in respiration. As a consequence, it can be used to compare virtually all eukaryote species.

Genetic diversity

REVISED

It is important to be able to measure **genetic diversity** within a species. A species with very little variation will struggle to adapt to any kind of environmental change. It is also important to be able to measure the genetic differences between different species. This gives us the basic information we require in order to construct evolutionary trees that show us what evolved from what.

There are several ways of measuring this genetic diversity:

1 **frequency of measurable or observable characteristics** — for example, spots on ladybirds or bands on snail shells
2 **base sequence of DNA** — different alleles have different base sequences
3 **base sequence of mRNA** — if there are different alleles, they will code for different mRNAs
4 **amino acid sequence** of the proteins encoded by DNA and mRNA

With **gene technology** (methods 2, 3 and 4), remember that the more the sequences are different, the more distantly related the organisms. This is because mutations accumulate over time.

Standard deviation and investigating variation within a species

REVISED

When investigating variation, it is not normally possible to look at every individual in a population. So, we have to look at a **sample**. We cannot get a realistic idea of the range of normal values unless we collect **random samples**, which means data collected without **bias**, i.e. without the conscious choice of the experimenter.

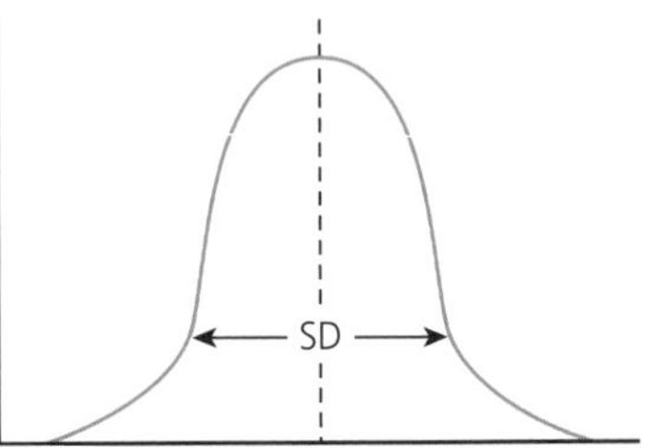

Figure 4.17 Standard deviation

When taking measurements, we want to get an understanding of what constitutes a normal range of values without our ideas being distorted by extreme individuals. **Standard deviation** expresses the spread of the data that is one-third either side of the **mean value** (Figure 4.17). One-third is 33.3%, which is normally rounded up to 34%, giving us a measure of the spread of the middle 68% of values.

For example, when measuring adult male height we could say that the standard deviation is 1.75 m ± 0.05, meaning that the mean height of adult males in our sample is 1.75 m and that 68% of them are between 1.7 m and 1.8 m tall.

Exam tip

Don't refer to *average* when you are referring to the *mean*. This is because median and mode are also types of average.

Exam tip

The AS specification clearly states that candidates will not be required to calculate standard deviations, but you might be asked to calculate the mean of a set of values.

Now test yourself

TESTED

20 Explain why standard deviation is not calculated using all of the measurements in a sample.

Answer on p. 205

Exam practice

1 If two organisms belong to the same family, to which other taxonomic groups do they also belong? [1]

2 These data were obtained from DNA hybridisation studies on great apes.

Species involved	Temperature at which hybrid DNA denatured (°C)
Human/human	94.0
Human/orangutan	90.4
Human/chimpanzee	92.4
Human/gorilla	91.7

(a) Explain why human/human DNA denatures at the highest temperature. [3]

(b) (i) Which species is the most distantly related to humans? [1]

(ii) Explain your answer to part (i). [2]

3 Cytochrome C is a commonly studied protein. It consists of 104 amino acids and is located in the mitochondria, where is has a vital function in respiration. The table shows the differences in the amino acid sequence between human cytochrome C and various other species.

Species pairings	Number of differences
Human/chimpanzee	0
Human/fruit fly	29
Human/horse	12
Human/rattlesnake	14
Human/rhesus monkey	1
Human/screw-worm fly	27
Human/snapping turtle	15
Human/tuna	21

(a) Why is cytochrome C a good protein to use in this type of study? [1]

(b) What is the minimum number of bases that must be contained in the gene for cytochrome C? [1]

(c) Explain why the actual number of bases is likely to be higher. [1]

(d) (i) Suggest which animal is the most distantly related animal to humans. [1]

(ii) Explain your answer to part (i). [3]

(e) Evaluate the statement: 'Humans are more distantly related to snapping turtles than to rattlesnakes.' [2]

4 A scientist studied the diversity of two areas of forest. The numbers of individuals of all species in the two habitats are given in the table below.

Species	Number of individuals in: Habitat A	Habitat B
A	34	87
B	30	2
C	25	1
D	30	1
E	15	1
F	0	1
G	0	1

(a) Explain what is meant by the term *random* sampling. [1]
(b) Which habitat shows the greatest species evenness? Explain your answer. [1]
(c) Use the following formula to calculate the diversity index of habitat B. [2]

$$d = \frac{N(N-1)}{\Sigma n(n-1)}$$

(d) Suggest three human activities that could result in a lowering of the index of diversity. [3]

5 Complete the table with a tick if each statement is correct or a cross if it is false. [3]

	Mitosis	Meiosis
DNA replicates		
Chromosome number is maintained		
Homologous chromosomes pair up		

6 The diagram shows the chromosomes in an organism where $2n = 6$.

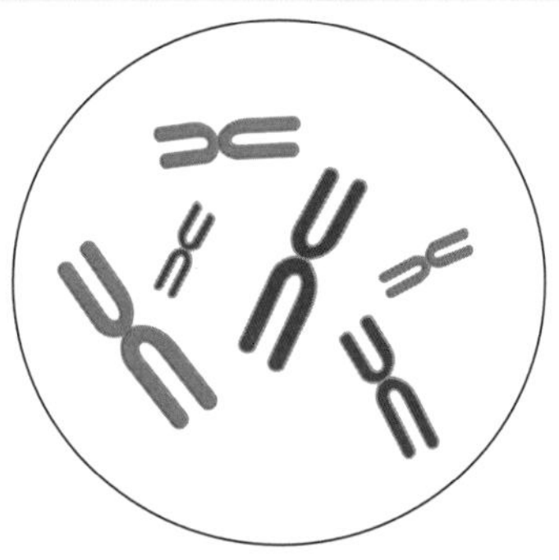

(a) Draw the chromosomes as they would appear after the first meiotic division. [2]
(b) Draw the chromosomes as they would appear after the second meiotic division. [2]

Answers and quick quiz 4 online

ONLINE

Summary

By the end of this chapter you should be able to understand:

- The distribution of DNA in prokaryotic and eukaryotic cells.
- The genetic code as a series of triplets.
- The distribution of coding and non-coding DNA in a cell.
- The processes of transcription and translation, including the roles of the different types of RNA.
- The ways in which meiosis can produce variation.
- The mechanism of natural selection and the two basic types of selection.
- The definition of a species in terms of observable similarities and the ability to produce fertile offspring.
- Courtship behaviour and how it can be used to distinguish different species.
- Phylogenetic groups are based on patterns of evolutionary history.
- Classification systems consist of a hierarchy in which groups are contained within larger composite groups and there is no overlap.
- One hierarchy comprises domain, kingdom, phylum, class, order, family, genus and species.
- Each species is universally identified by a binomial name.
- Biodiversity can be measured by an index of diversity that involves the number of different species and the number of individuals of each species.
- Farming techniques can reduce biodiversity.
- Evolutionary differences can be assessed by looking at base sequences in DNA and amino acid sequences in proteins.
- Standard deviation is a measure of the spread of the data about the mean.

5 Energy transfers in and between organisms

Photosynthesis

Photosynthesis is vital to life on Earth because it:

- is the only way that energy can get into ecosystems
- turns a colourless, odourless gas (carbon dioxide) into organic molecules such as glucose and starch
- creates oxygen as a by-product

Chloroplasts are the organelles of photosynthesis (Figure 5.1). They give the green colour to all upper parts of a plant, but they are mainly concentrated in the **palisade cells** of the leaves. Chloroplasts were studied in Chapter 2. The key features are:

- a light-harvesting group of compounds collectively called **chlorophyll**
- the chlorophyll is embedded in flat discs called **thylakoids**
- the thylakoids are packed into stacks called **grana** (singular: granum)
- between the grana is an enzyme-rich fluid called the **stroma**

Photosynthesis is split into two key steps: the light-dependent reaction and light-independent reaction.

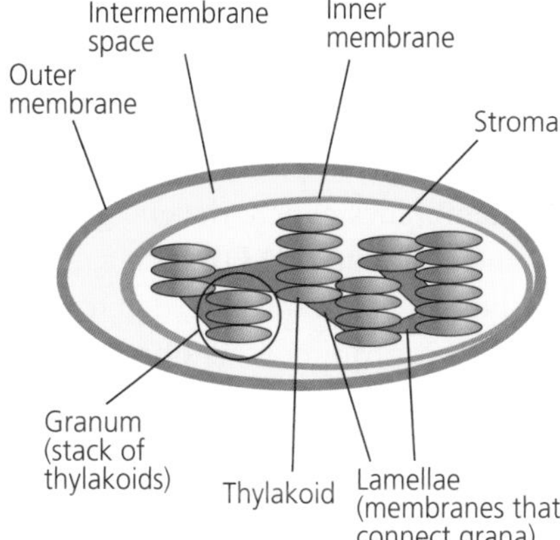

Figure 5.1 The structure of a chloroplast

Measuring the rate of photosynthesis

REVISED

Photosynthesis is usually measured using an aquatic plant — often *Elodea* or *Cabomba*. This is because the oxygen is given off as bubbles that can be collected easily (Figure 5.2).

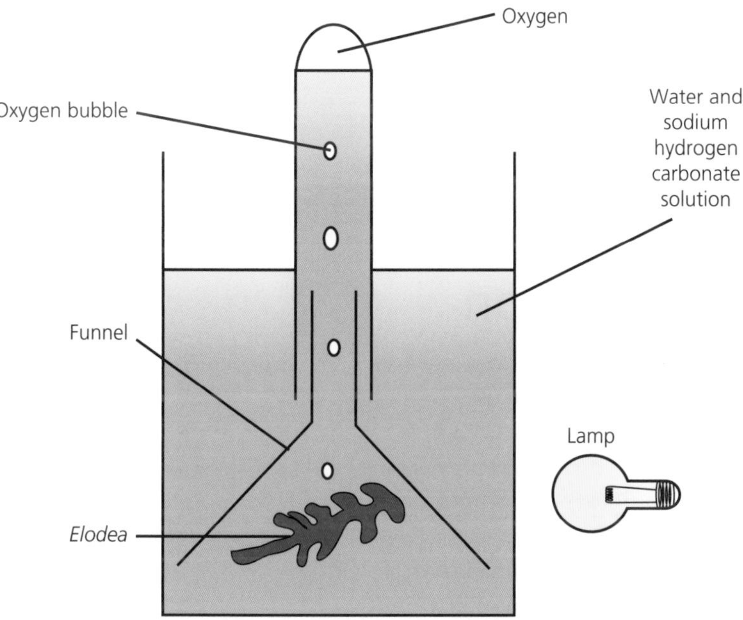

Figure 5.2 Measuring photosynthesis using an aquatic plant

Using this apparatus you can investigate the effect of various limiting factors on the rate of photosynthesis:

- Aquatic plants get their carbon dioxide in solution, as HCO_3^- ions. Therefore, carbon dioxide levels can be varied by using different concentrations of HCO_3^-.
- Light intensity can be varied by placing a lamp at different distances from the plant. Light intensity is proportional to $\frac{1}{d^2}$ where d = distance. This means that if you double the distance, the light intensity will be one-quarter of the original value.
- Temperature can be varied by heating or cooling the water.

The light-dependent reaction

This first process relies on chlorophyll, so it takes place in the thylakoids. The key steps are as follows:

- Chlorophyll absorbs light and emits **excited** electrons. This process is called **photoionisation**.
- The electrons pass along an electron transport chain in the thylakoid membranes, where the energy is used to make two vital compounds: **ATP** and **reduced NADP** (also called **NADPH**). They possess the energy and the reducing power needed to turn carbon dioxide into glucose.
- ATP is produced via **chemiosmosis**. Energy in the electrons is used to pump H^+ ions (protons) from the stroma into the thylakoid membrane. This results in a diffusion gradient. The protons diffuse back into the stroma through the enzyme ATP synthase, which makes ATP from ADP and P_i. This is very similar to the way ATP is made in respiration (page XX).
- In order for photosynthesis to continue, chlorophyll needs replacement electrons. These come from splitting water in a process called **photolysis**.
- Photolysis creates electrons, hydrogen ions and oxygen gas as a by-product. That's where the oxygen in our atmosphere originates.

Excited in this context means raised to a higher energy level.

Exam tip

When studying this topic, stick to the specification. You don't need anything more than the details above. If you come across photosystem 1 and photosystem 2, you are going into too much detail.

Now test yourself

TESTED

1 Apart from light, list four things that are needed for the light-dependent reaction.
2 In the light-dependent reaction, name the two compounds that are necessary in order for the plant to reduce carbon dioxide to glucose.
3 Explain how photosynthesis produces oxygen.

Answers on p. 205

The light-independent reaction

The light-independent reaction takes place in the stroma of the chloroplasts and involves the reduction of carbon dioxide to glucose. These reactions are sometimes called the **Calvin cycle** (Figure 5.3). The key steps are as follows:

- One molecule of carbon dioxide combines with a 5-carbon compound called **ribulose bisphosphate (RuBP)**.
- This results in a 6-carbon compound that immediately splits into two molecules of **glycerate 3-phosphate (GP)**. The step is catalysed by the enzyme **rubisco.** GP has three carbon atoms but is not a sugar.
- ATP and reduced NADP (from the light-dependent reaction) bring about reduction of GP to **triose phosphate (TP)**, which is the first sugar.
- Some triose phosphate is eventually turned into **glucose**, and then into other useful organic compounds such as starch. Some triose phosphate is used to resynthesise RuBP.

- Every cycle of reactions increases the number of carbon atoms by one — the one in carbon dioxide. So it takes six cycles to accumulate enough carbon to make one molecule of glucose. Each cycle also regenerates RuBP so that the cycle can continue.

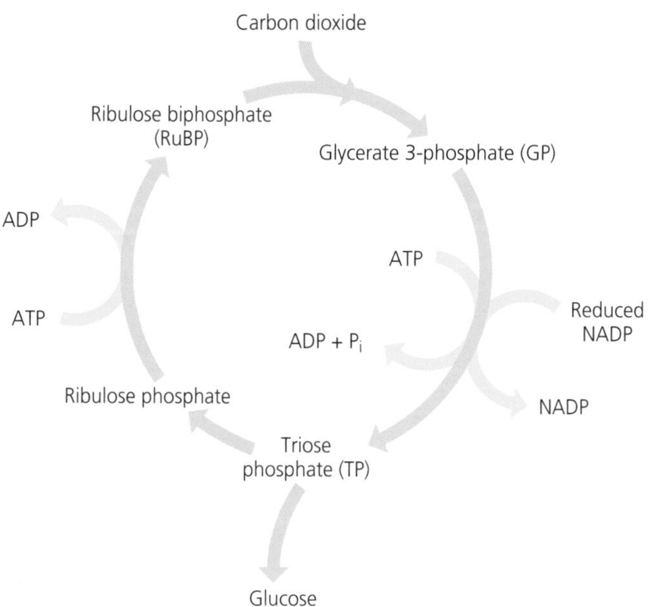

Figure 5.3 A simple Calvin cycle

Exam tip

The specification states that 'triose phosphate is converted to useful organic substances', which means glucose, starch, sucrose and various other organic molecules. Plants also synthesise proteins, phospholipids and nucleic acids using the mineral ions — notably nitrate and phosphate — absorbed by the roots.

Exam tip

The specification refers to photosynthesis in a typical C3 plant, which means that the first sugar made (triose phosphate) has three carbons. Most plants are C3 plants; there are C4 plants, but you don't need to know about them.

Revision activity

Make sure you can draw a diagram to show the light-independent reaction, showing the three intermediates (RuBP, GP and TP) and the role of ATP and reduced NADP.

Typical mistake

Photosynthesis and respiration often feature in long exam questions, for 5 or 6 marks. No matter how carefully these are worded, students often interpret them as 'write out everything you know'. Read the question carefully. You will be told exactly which part to write about, such as the light-dependent reaction.

Limiting factors

What limits the rate of photosynthesis?

REVISED

The definition of a limiting factor is one which, if the supply is increased, will speed up the process. The limiting factors for photosynthesis are:

- **temperature** — the light-independent reaction is temperature-sensitive because it is enzyme-controlled; the light-dependent reaction is less so
- **carbon dioxide** — needed as the source of carbon
- **light intensity** — needed to excite the chlorophyll

Knowing about limiting factors is important if you want to make plants grow faster — a basic requirement of agriculture.

Generally, increasing the amount of light will increase the rate of photosynthesis until carbon dioxide becomes limiting. Then, increasing the amount of carbon dioxide will increase the rate until temperature becomes limiting. When the temperature is at an optimum the plant will be photosynthesising at its maximum rate, which is then just limited by the amount of chlorophyll it possesses.

For maximum growth, plants also need a supply of **mineral ions** — the key ones are **nitrate**, **phosphate** and **potassium**. That is why fertiliser is needed. The application of excess fertiliser can cause problems, such as eutrophication (page XX).

Respiration

ATP synthesis

REVISED

Respiration is one of the seven signs of life. Its purpose is to transfer the energy from organic molecules, such as glucose and lipid, into ATP. There are two types of respiration:

- **aerobic respiration**, which requires oxygen
- **anaerobic respiration**, which does not require oxygen

Aerobic respiration takes place in four stages: **glycolysis**, which takes place in the cytoplasm, followed by the **link reaction**, **Krebs cycle** and **electron transport chain**, all of which take place in the **mitochondria**.

Anaerobic respiration is basically glycolysis that cannot go any further, usually because there is no oxygen.

Exam tip

Make sure you know the structure and function of the ATP molecule, covered in Chapter 1, page XX.)

Exam tip

Don't say that 'respiration *makes* energy', because it doesn't. Respiration *releases* the energy stored in organic molecules and *transfers* it into ATP. 'Transfer' or 'release' are good words. 'Makes' or 'creates' is a crime against physics — you will have broken the first law of thermodynamics, which says that energy cannot be made or destroyed.

Now test yourself

TESTED

4 Of the four processes in aerobic respiration, list all those that make ATP.

Answer on p. 205

Mitochondria

REVISED

Mitochondria (Figure 5.4) are the organelles of aerobic respiration. They are found in eukaryotic cells, where they are the source of most of the cell's ATP. A mitochondrion is bounded by two **membranes**: the outer membrane, which is freely permeable to many substances, and the inner membrane, which is more selectively permeable. Most substances can only pass through the inner membrane via appropriate carrier molecules. The inner membrane is also the site of the molecules of the **electron transport chain** and membrane-bound ATP synthase enzymes. It is folded into **cristae**, which increases the area of the membrane and allows increased electron transport and ATP synthesis.

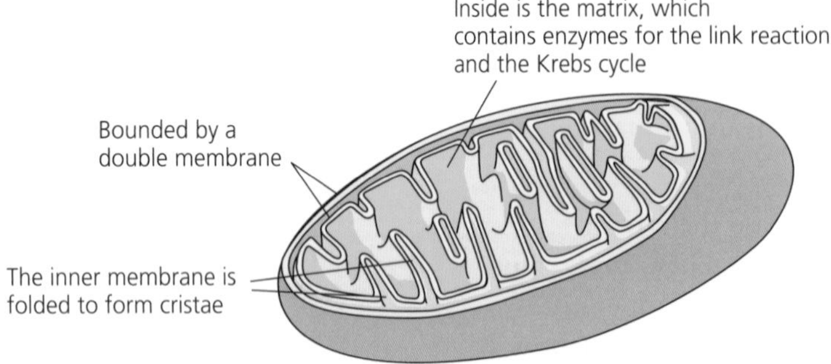
Inside is the matrix, which contains enzymes for the link reaction and the Krebs cycle
Bounded by a double membrane
The inner membrane is folded to form cristae

Figure 5.4 A mitochondrion. Full aerobic respiration consists of four processes, the last three of which take place in the mitochondria. The link reaction and the Krebs cycle take place in the fluid matrix in the centre, whereas the electron transport chain takes place on the inner membrane

Typical mistake

Students often assume that mitochondria need, absorb or use glucose. They do not. Glucose is used up in glycolysis, which happens in the cytoplasm. Mitochondria absorb the pyruvate that results from glycolysis.

Glycolysis

REVISED

Glycolysis means 'sugar splitting' and it is a universal process — it happens in the cytoplasm of all cells. Aerobic and anaerobic respiration both start with glycolysis. The key features are as follows:

- Glucose is **phosphorylated** — it is raised to a higher energy level using two molecules of ATP.
- The resulting molecule is split into two molecules of **triose phosphate** (a 3-carbon compound).
- Triose phosphate is split into two molecules of **pyruvate**, also a 3-carbon compound.
- The splitting of pyruvate creates two vital compounds: four molecules of ATP, giving a net profit of two; and two molecules of reduced coenzyme, NADH.

What happens next — either aerobic or anaerobic respiration — depends on whether or not oxygen is available.

> **Exam tip**
>
> Glycolysis consists of 10 enzyme-controlled steps. The enzymes for glycolysis are found in the cytoplasm of virtually all cells in all organisms. This is indirect but very clear evidence for evolution — all organisms share a common ancestor.

Aerobic respiration

The link reaction

REVISED

If oxygen is available, and it is a eukaryotic cell, pyruvate passes into the mitochondria by active transport where it enters the link reaction, also called **pyruvate oxidation**. In the matrix of the mitochondria, pyruvate is split to form acetate (a 2-carbon compound) and carbon dioxide. The acetate combines with a larger molecule, **coenzyme A**, to produce **acetylcoenzyme A** (sometimes just called acetyl coA). This process produces no ATP but it does make NADH.

The Krebs cycle

REVISED

The Krebs cycle is a series of oxidation and reduction reactions whose purpose is to remove electrons (and hence energy) from what is left of the glucose, now just a 2-carbon acetate. It begins when acetyl coA combines with a 4-carbon compound to become a 6-carbon compound (Figure 5.5). The 6-carbon compound enters a series of reactions in the matrix, which produces:

- carbon dioxide
- ATP from substrate-level phosphorylation
- reduced coenzymes NADH and $FADH_2$
- more 4-carbon compounds to continue the cycle

The cycle turns once for every molecule of acetyl coenzyme A, and so it turns twice for every glucose molecule. The Krebs cycle can also be powered by the breakdown products of lipids and amino acids, allowing organisms to respire aerobically, and keep making ATP, even when no glucose is available. In many carnivores, for example, a lot of their ATP comes from the respiration of amino acids from their high-protein diet.

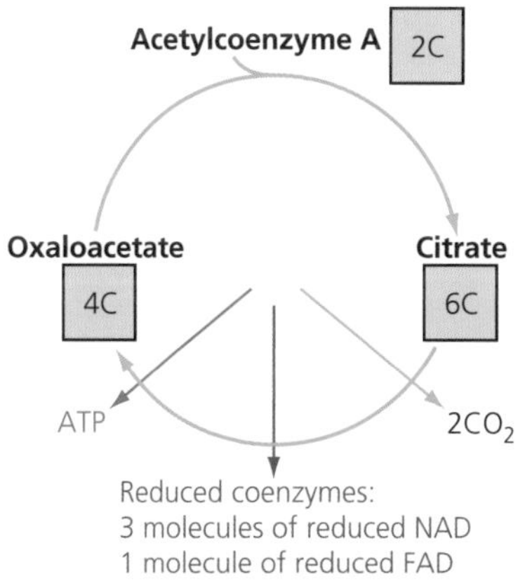

Figure 5.5 The essentials of the Krebs cycle. The cycle turns twice for each glucose molecule

> **Exam tip**
>
> $FADH_2$ is similar to NADH and does pretty much the same electron-carrying function.

> **Exam tip**
>
> You don't need to know the names of the compounds involved in the Krebs cycle.

The electron transport chain

REVISED

> **Oxidative phosphorylation** is the part of the respiratory pathway in which energy released in the electron transfer chain is used in the production of ATP.

This process makes a lot of ATP by the process of **oxidative phosphorylation** and takes place on the inner mitochondrial membrane (Figure 5.6). It involves all the reduced coenzymes made by the first three processes — the energy in the electrons they carry is used to make ATP:

- Reduced coenzymes deliver their electrons to the proteins of the electron transport chain.
- The electrons pass along a series of proteins, which are first reduced and then oxidised. These redox reactions release energy that is used to pump **protons** (or H^+ ions) into the outer mitochondrial space. The result is a high concentration of H^+ ions in the outer mitochondrial space.
- The protons diffuse back into the matrix through the middle of **ATP synthase** enzymes. As they pass through the enzymes, ATP is synthesised. This is known as **chemiosmosis**.

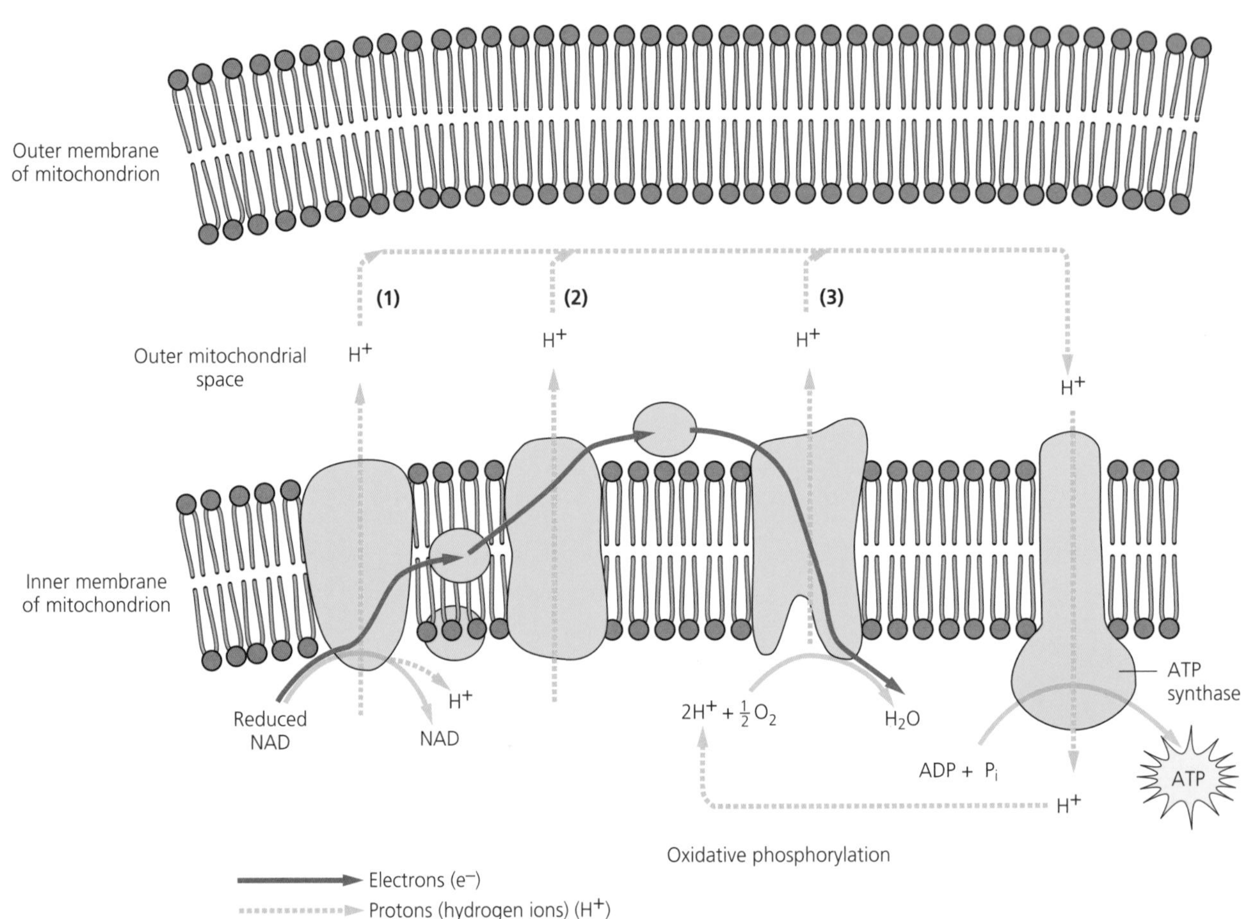

Figure 5.6 ATP is made by oxidative phosphorylation on the inner mitochondrial membrane

Why is oxygen so important?

REVISED

The end product of the process is low-energy electrons, which need to be mopped up. This is why oxygen is needed. The oxygen combines with electrons and hydrogen ions to form water:

$$4H^+ + 4e^- + O_2 \rightarrow 2H_2O$$

If oxygen is in short supply, there is nothing to accept the electrons at the end of the process, so the electron transport chain stops and no ATP is produced. This is fatal for most organisms in a very short period of time.

Phosphorylation

REVISED

There are two types of **phosphorylation**. ATP is made in glycolysis and the Krebs cycle by **substrate-level phosphorylation**, which means that the phosphate used to make ATP comes from a substrate, i.e. another molecule. The other way to make ATP is by **oxidative phosphorylation**, which takes place in the electron transport chain. Here, ATP is made by a series of electron transfers followed by the diffusion of hydrogen ions across the mitochondrial membrane.

Phosphorylation simply means 'adding a phosphate'.

Substrate-level phosphorylation results directly in the formation of ATP from ADP and phosphate. It does not involve oxidative phosphorylation. In aerobic respiration, substrate-level phosphorylation occurs in glycolysis and the Krebs cycle.

Exam tip

If a question says 'Describe how oxidative phosphorylation makes ATP', it effectively means 'Tell us about the electron transport chain'.

Now test yourself

TESTED

5 Explain what is meant by:
(a) substrate-level phosphorylation
(b) oxidative phosphorylation

Answer on p. 205

How much ATP is made in aerobic respiration?

REVISED

So far, for every glucose molecule, we have:

- 2 ATP molecules from glycolysis, which is all that anaerobic respiration produces
- 2 ATP molecules from the Krebs cycle

That's it for substrate-level phosphorylation. However, there are many reduced coenzymes carrying electrons that are used to power oxidative phosphorylation. These are:

- 2 NADH molecules from glycolysis
- 2 NADH molecules from the link reaction
- 6 NADH molecules from the Krebs cycle
- 2 $FADH_2$ molecules from the Krebs cycle

When they deliver their electrons to the inner mitochondrial membrane, each NADH produces 3 ATP molecules and each FADH produces 2 ATP molecules. Adding it all up, there are 10 NADH providing 30 ATP molecules and 2 $FADH_2$ providing 4 ATP molecules. So the grand total is 38 ATP molecules, 34 of which come from oxidative phosphorylation.

Figure 5.7 summarises the main processes in respiration.

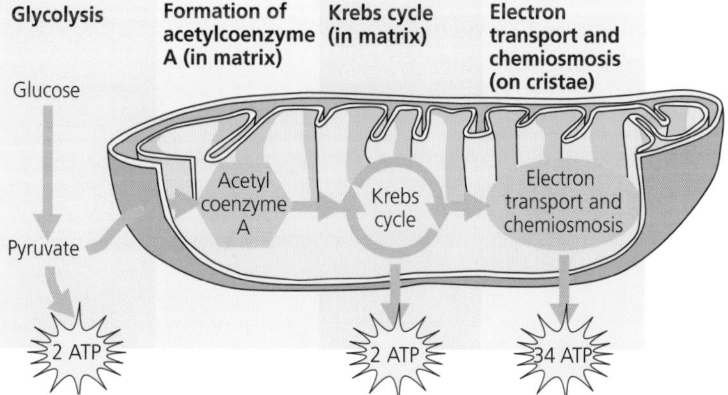

Figure 5.7 Summary of the main processes in respiration

Now test yourself

TESTED

6 Explain how the structure of mitochondria maximises the process of oxidative phosphorylation.

7 For each glucose molecule, state how many ATP molecules are made in:
(a) anaerobic respiration
(b) aerobic respiration

Answer on p. 205

Exam tip

Don't worry too much about learning the exact amounts of NADH and ATP at each stage. The vital point is that oxidative phosphorylation, which is the key process at the end of aerobic respiration, produces far more ATP than any other process.

Anaerobic respiration

What is anaerobic respiration?

REVISED

In the absence of oxygen, respiration cannot continue because pyruvate cannot be oxidised further. The cell or organism must survive on the small amount of ATP made in glycolysis. The problem is that glycolysis cannot continue when all of the coenzyme has been reduced to produce NADH. The key step in anaerobic respiration is to reduce the pyruvate *so that the coenzyme NAD$^+$ is resynthesised*:

- In animals and many bacteria, the pyruvate is converted into **lactate**.
- In plants and many fungi (including yeast), the pyruvate is converted into **ethanol** (alcohol) and carbon dioxide — the basis of alcoholic fermentation.

Typical mistake

Students often think that anaerobic respiration is a completely different process from aerobic respiration, when in reality anaerobic respiration is just glycolysis that cannot go any further.

Now test yourself

TESTED

8 Is NAD$^+$ reduced or not? Explain your answer.

Answer on p. 205

Measuring respiration

Using a respirometer

REVISED

Respiration can be measured by oxygen uptake using a respirometer like the one in Figure 5.8. The organism respires, taking in oxygen and giving out carbon dioxide. Normally, this would not change the volume in the chamber, but the sodium hydroxide absorbs all the carbon dioxide. So it is just as if the organism is not making carbon dioxide at all. As a result, the volume in the chamber reduces as the organism uses oxygen. You can measure how much oxygen is being used by reading the scale.

To compare different organisms, three measurements are needed:

- the mass of the organisms
- the volume of oxygen used
- the time taken

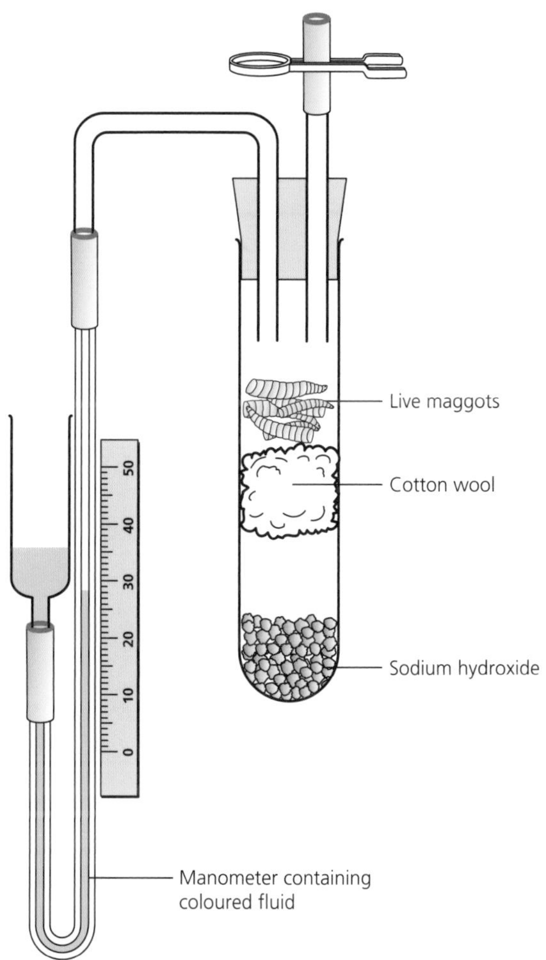

Figure 5.8 A simple respirometer

Now test yourself

TESTED

9 Suggest suitable units for measuring respiration.

Answer on p. 205

Example

Interpreting data

1 Study the following graph, which shows the rates of photosynthesis and respiration in a crop that is normally grown in a glasshouse.

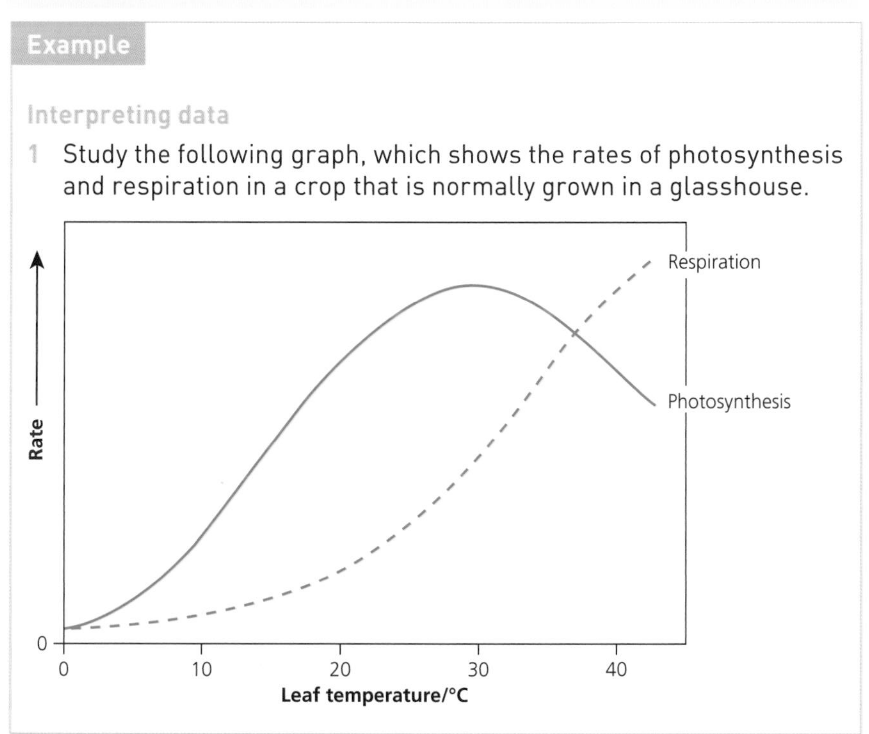

(a) Explain why respiration and photosynthesis are temperature sensitive.
(b) Explain how the rate of photosynthesis affects crop yield.
(c) Explain how the rate of respiration affects crop yield.
(d) Predict the temperature at which crop yield is greatest. Explain your answer.
(e) Predict the effect of an increase in the temperature of the climate from the optimum to 3°C higher.

Answer

(a) They are enzyme-controlled processes. (To be specific, the light-independent reaction of photosynthesis and the whole of respiration are enzyme-controlled and therefore temperature-sensitive. The light-dependent stage of photosynthesis, involving the excitation of chlorophyll, is not directly enzyme-dependent.)
(b) The higher the rate, the greater the yield (it makes organic molecules).
(c) The higher the rate, the lower the yield (it uses organic molecules).
(d) About 25°C — this is the temperature at which photosynthesis exceeds respiration by the most.
(e) Loss in yield because of higher respiration.

Exam practice

1 Complete the following table. [3]

Question	Photosynthesis	Respiration
Is ATP made?		
Is ATP made in an electron transport chain?		
Is ATP made by substrate-level phosphorylation?		

2 Explain how the structure of the ATP molecule allows it to perform its role in the cell. [4]

Answers and quick quiz 5 online

ONLINE

Summary

By the end of this chapter you should be able to understand the following:

- Photosynthesis has two key stages: the light-dependent and the light-independent reactions.
- The light-dependent reaction produces ATP and NADPH.
- The light-independent reaction uses ATP and NADPH to convert carbon dioxide into glucose.
- Temperature, carbon dioxide concentration and light intensity are the main limiting factors for the rate of photosynthesis.
- Full aerobic respiration has four stages: glycolysis, the link reaction, the Krebs cycle and the electron transport chain.
- In glycolysis, glucose is converted into pyruvate.
- In the link reaction, pyruvate is converted into acetylcoenzyme A.
- In the Krebs cycle, the electrons in the acetate are used to make ATP and reduced coenzymes.
- In the electron transport chain, the electrons carried by the reduced coenzymes are used to make ATP by oxidative phosphorylation.
- Anaerobic respiration consists of glycolysis followed by a simple step that resynthesises NAD^+ from NADH, so that glycolysis can continue.

6 Energy and ecosystems

Net productivity of producers

The foundation of any ecosystem is the photosynthetic activity of the **producers**. They take CO_2 from the air — or, if aquatic, HCO_3^- ions from the water — and use it as the raw material for making organic molecules.

Producers include **plants**, **algae** and **some bacteria**. They capture sunlight energy and incorporate it into the chemical bonds of the carbohydrates they synthesise. Some of this energy is used by the plants: remember that plants respire too. The energy available to the rest of the food chain is the energy locked in the starch, cellulose, sucrose and other substances when the plant is eaten. These materials make up the **biomass** of the crop.

This gives us two key terms:

Gross primary productivity (GPP) — which refers to the energy in all the organic molecules made by photosynthesis. Think of it as your total salary.

Net primary productivity (NPP) — which refers to the amount of energy left once the plant has used what it needs in respiration. Think of it as your take-home pay after tax.

An important equation is:

net productivity = gross productivity – respiratory loss

Plants require energy for their own metabolic processes, just like animals. NPP is a measure of how much the process of photosynthesis exceeds respiration. This is the energy that accumulates in the plant, and that's what is available to an animal when it eats the plant. To a farmer NPP represents the crop, i.e. what they can sell on.

Biomass and calorimetry

REVISED

Biomass can be measured in terms of mass of carbon or, more practically, the dry mass of tissue per given area per given time. If you wanted to study the productivity of an ecosystem or a crop, for example, you could measure biomass in units such as $kJ\,m^{-2}\,d^{-1}$, which means kilojoules per square metre per day.

How do you estimate the energy content of living or dead tissue? The simple answer is to dry it, burn it in oxygen and measure the energy given off. This process is called calorimetry (Figure 6.1). To estimate the biomass and energy content of, say, a grassland, you would take all of the plant material in one square metre and dry it in an oven at 80°C to remove the water. Then you could burn the dry material in a calorimeter to estimate the energy content.

Now test yourself

TESTED

1 Biomass is measured in units such as $kJ\,m^{-2}\,d^{-1}$. Explain why it is important to include the 'per unit area' and 'per unit time'.

Answer on p. 205

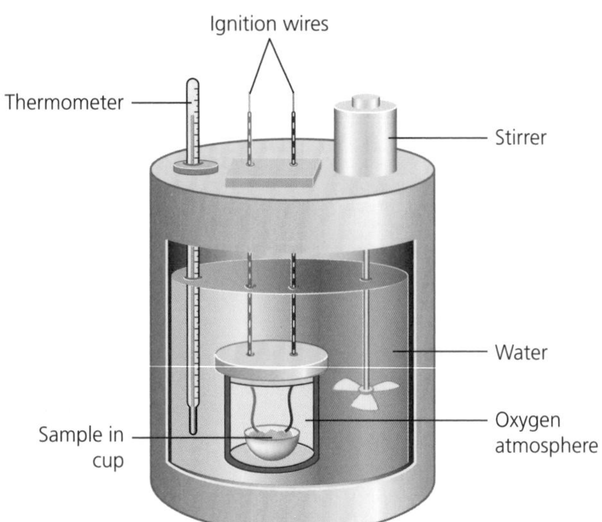

Figure 6.1 **A calorimeter. This is how the energy content for food labels is estimated. A dry sample of the material is burned in pure oxygen and the energy released heats up the surrounding water**

Typical mistake

Confusing **calorimetry**, which is the measurement of the energy content, with **colorimetry**, which is the measurement of the absorbance or transmission of a liquid. You probably used a colorimeter in year 1.

Net productivity of consumers

In terms of efficiency, the simpler the food chain, the less energy is lost. It is very wasteful to produce meat — it is far more efficient to grow and eat plants because there is just one link in the food chain. However, the demand for meat remains strong, so it is *economically* viable to produce meat; just wasteful in terms of energy.

The net productivity (N) of consumers such as animals can be calculated from the equation:

$$N = I - (F + R)$$

Where:

I = the energy in the ingested food

F = the energy lost in the faeces

R = the energy lost in respiration

This is important in farming. A farmer will want net productivity to be as high as possible, because that represents meat that can be sold. There are several ways to achieve this:

- Feed concentrates so that more food is more digestible. This reduces the energy lost in faeces. A lot of the dry mass of foodstuffs such as grass is cellulose, which is not easy to digest.
- Reduce the energy lost via respiration by keeping the animals in heated barns and/or restricting their movement.
- Selective breeding for rapid growth, or other desirable characteristics such as meat/fat ratio.

Now test yourself TESTED

2 Explain how keeping animals in heated barns reduces the energy lost via respiration.

Answer on p. 205

Nutrient cycles

The role of microorganisms

REVISED

The atoms and molecules that make up our bodies have all been part of many other organisms before and will be part of many more in the future. Remember that elements are recycled, but energy is not. Energy from the Sun drives the cycles. If the Sun stopped shining, photosynthesis would stop and so would all life on Earth.

Nutrient cycles rely on the action of **saprobionts** (bacteria and fungi) in making the elements locked up in complex organic molecules available to plants once again. These saprobionts feed by **extracellular digestion**, which breaks down dead organic matter. They synthesise and secrete digestive enzymes on to the surface of their food. These enzymes break down complex molecules such as DNA and protein, releasing soluble ions such as **nitrate** and **phosphate**.

Exam tip

Language is important in exams, and the examiners want to see evidence of what you have learned at A-level. The terms **rotters, decomposers** and **saprobionts** all refer to the same thing, but the examiners will credit use of the word saprobionts. That's proper A-level terminology.

The nitrogen cycle

REVISED

Nitrogen is an essential component of proteins and nucleic acids (DNA and RNA), along with smaller molecules such as ATP, urea and ammonia. Some 80% of air is nitrogen gas (N_2), but the molecule is stable and not usually available to organisms. It takes a lot of energy to split molecules of nitrogen gas because the two atoms are connected by a strong triple bond.

Now test yourself TESTED

3 Explain why atmospheric nitrogen is not normally available to organisms.

Answer on p. 205

Plants can only absorb nitrogen as **nitrate** ions (NO_3^-). These ions are present in the water in soil and are absorbed into the roots by **active transport**. Plants combine nitrate with the substances made in photosynthesis to make amino acids and nucleotides that, in turn, make proteins and nucleic acids.

In the nitrogen cycle (Figure 6.2), nitrogen passes up the food chain in organic molecules. Animals get their nitrogen in proteins and nucleic acids. Eventually, all the nitrogen ends up in dead organic matter: dead plants and animals, faeces and urine. At this stage, the nitrogen is still locked up in organic molecules, so the action of saprobionts (bacteria and fungi) takes over. **Saprobiont nutrition** involves the breakdown of large molecules by extracellular digestion. Proteins are broken down into amino acids, which the saprobionts absorb. These amino acids are **deaminated** and the resulting ammonium (NH_4^+) ions are released as a by-product in a process called **ammonification**.

Deamination is the removal of the amino (NH_2) group from an amino acid.

Ammonification is the stage in the nitrogen cycle in which saprobiotic microorganisms break down organic, nitrogen-containing substances such as proteins and produce ammonium compounds.

The next stage is **nitrification**. There are two types of **nitrifying bacteria**, each of which plays a part in converting ammonium into nitrate in a two-stage process: first, ammonium is oxidised into nitrite ions; then this nitrite is further oxidised into nitrate ions. Plants can absorb the nitrate and the cycle continues.

Nitrification is an important stage in the nitrogen cycle in which ammonium compounds are converted to nitrites and nitrates.

However, there are two complications to the nitrogen cycle:

- **Nitrogen fixation** — nitrogen-fixing bacteria contain the enzyme **nitrogenase**, which converts nitrogen gas into ammonium ions so that nitrogen can re-enter the cycle. These bacteria can be found free-living in the soil, or in the root nodules of leguminous plants such as peas and beans. Nitrogen gas can also be fixed by lightning during electrical storms.
- **Denitrification** — denitrifying bacteria lose nitrate from the cycle by converting nitrate ions into nitrogen gas. This process tends to occur in waterlogged, anaerobic soil.

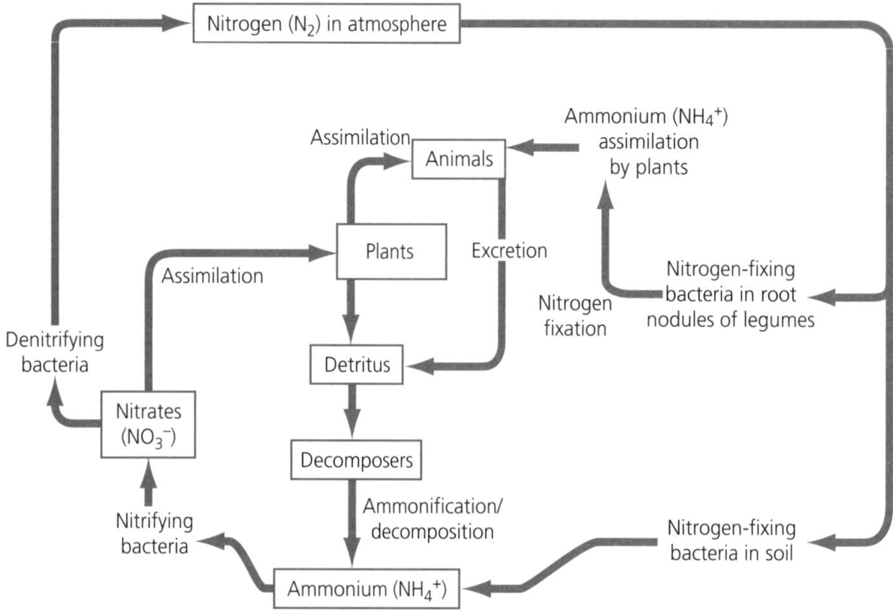

Figure 6.2 The main stages in the nitrogen cycle

Now test yourself

TESTED

4 Explain what is meant by the term *nitrogen fixation*.
5 List the *two* ways in which nitrogen can be fixed.
6 Some plant species, such as clover, beans and peas (legumes), have root nodules that contain nitrogen-fixing bacteria. Suggest the advantage of this to:
  (a) the plant
  (b) the bacteria
7 Nitrogen-fixing bacteria have an enzyme called nitrogenase, which has the ability to fix nitrogen gas into ammonia. Suggest why scientists want to put this enzyme into other agricultural crops such as wheat and corn.

Answers on pp. 205–6

Exam tip

The specification is very clear — you do *not* need to know the names of any of the bacteria in the nitrogen cycle. If your notes are covered with names like *Nitrosomonas* and *Nitrobacter*, delete them.

The role of mycorrhizae

Many plants — including most tree species — have mutualistic arrangements with various species of fungi, collectively called mycorrhizae (literally meaning 'fungus-root'). These fungi grow in and around the root system, increasing the surface area for the absorption of water and inorganic ions (Figure 6.3). In return, the plant provides the fungus with sugars from photosynthesis.

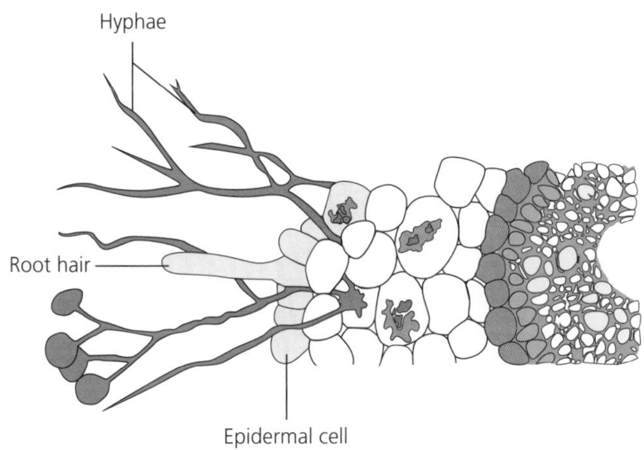

Figure 6.3 With mycorrhizae, the fungal threads extend into the soil and increase the surface area for absorption

The phosphorus cycle

REVISED

Phosphorus is an essential component of DNA, RNA, ATP, phospholipids and some proteins. The essential steps of the phosphorus cycle (Figure 6.4) are as follows:

1 The action of saprophytes releases phosphate ions (PO_4^-) in much the same way as for nitrate ions (see above).
2 Plant roots absorb phosphate ions by active transport and incorporate them into the organic molecules mentioned above.
3 Phosphorus passes up the food chain in organic molecules. Eventually phosphorus ends up in dead organic matter where the action of saprophytes releases phosphate ions, bringing the cycle full circle.

Exam tip

A key difference between the P cycle and the N cycle is that the P cycle doesn't normally involve the atmosphere.

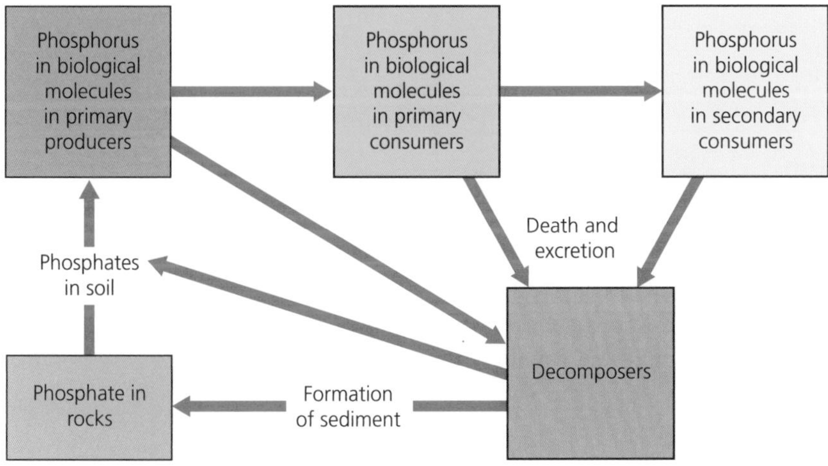

Figure 6.4 The phosphorus cycle

Exam tip

There is a geological aspect to the phosphorus cycle. Phosphate can be added to the cycle by the weathering of certain minerals such as apatite. However, it is unlikely that this aspect will be included in biology exams.

Leaching and eutrophication

REVISED

Leaching is the loss of soluble substances such as nitrates and phosphates from the top layer of the soil when water drains through. When farmers apply **fertilisers** to their land, they often get washed off by rainfall and find their way into rivers or lakes, where they can cause **eutrophication**. The nitrate, phosphate and other ions over-fertilise the aquatic ecosystem and cause the following process:

1. The ions cause a bloom of algae because these reproduce faster than aquatic plants.
2. The algae block the light, causing the death of the plants below.
3. When the algae have used up all the nutrients they also die, adding to the dead matter.
4. Saprobiotic decay by aerobic bacteria uses up all the oxygen.
5. A 'dead zone' results and all the native organisms that require oxygen die.

Eutrophication is an increase in the quantity of plant nutrients. The term is used when freshwater lakes or rivers are enriched with nitrates and phosphates, either as a result of the leaching of fertiliser from agricultural land or from sewage effluent.

Revision activity

Draw a flow diagram to summarise the stages involved in eutrophication.

Now test yourself

TESTED

8 What evidence is there in the text that mineral ions are limiting factors for algal growth?

Answer on p. 206

Exam practice

1 The following diagram shows the nitrogen cycle.

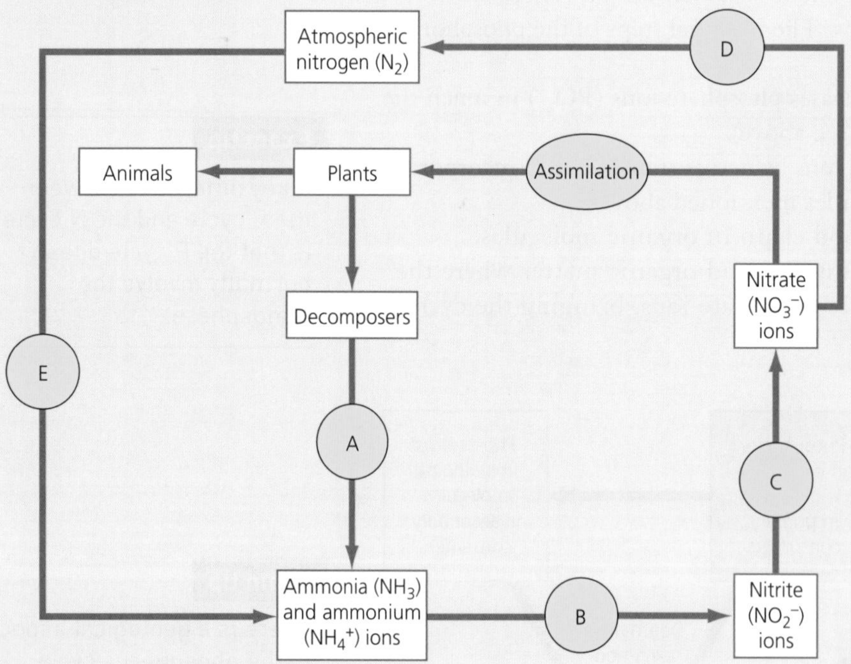

(a) Name a compound that is classed as 'organic nitrogen'. [1]

(b) Name processes A, B, C, D and E. [2]

(c) Suggest why many crop rotation schemes include growing a legume such as beans. [3]

2 The following table shows the average dates of the first spawning of frogs in ponds along the south coast of England.

Year	Average date of first spawning
1955	16 February
1960	17 February
1965	13 February
1970	9 February
1975	7 February
1980	3 February
1985	1 February
1990	27 January
1995	23 January
2000	18 January

(a) Describe the pattern shown by the data in the table. [1]
(b) Suggest how the data would be different for ponds in Scotland. Explain your answer. [2]
(c) Suggest how the changing pattern may be a problem for the frog population. [2]
(d) Animals and plants that use temperature to synchronise their life cycles are affected by global warming. Suggest a more reliable environmental factor. [1]

Answers and quick quiz 6 online

ONLINE

Summary

By the end of this chapter you should be able to understand the following:

- Energy enters an ecosystem when it is captured in photosynthesis, although a large percentage of sunlight is lost. Some is reflected off leaves, some misses the chloroplasts and some is the wrong wavelength.
- All energy transfers are inefficient, so energy is lost at each trophic level. The more steps in the food chain, the more energy is lost.
- Gross primary production is basically the sum total of the organic molecules made in photosynthesis. Net primary production is what remains when respiratory loss is taken into account.
- Farming practices increase the efficiency of energy conversion. In terms of energy it is much more efficient to grow plants than to produce meat.
- Measures to increase the efficiency of meat production include selective breeding, minimising movement and heat loss, and feeding concentrates so that less energy is lost in undigestible material such as cellulose.
- Microorganisms have an important role in the nitrogen and phosphorus cycles. Both cycles involve the basic process of saprobiotic nutrition (decomposition), which involves extracellular digestion. This process releases ammonium and phosphate ions.
- In the process of nitrification, nitrifying bacteria oxidise ammonium ions into nitrite and then nitrate — the form of nitrogen that plants can absorb.
- Nitrogen is added to the ecosystem by the process of nitrogen fixation, in which N_2 gas is converted into ammonium ions. This can be done by electrical storms but most nitrogen is fixed by nitrogen-fixing bacteria.
- Nitrogen is lost from the ecosystem by the process of denitrification, in which bacteria convert nitrate into nitrogen gas.
- The use of fertilisers is widespread because mineral ions such as nitrate, phosphate and potassium are limiting factors in crop growth.
- There are serious environmental issues arising from the use of fertilisers, including leaching and eutrophication.

7 Stimulus and response

Survival and response

All organisms have the same aim in life: to try to get food and to avoid being food for something else until they have had a chance to reproduce and pass on their alleles. It is vital that organisms are able to detect and respond to changing environmental conditions. You may remember that sensitivity is one of the seven signs of life.

A **stimulus** is a change in the environment that can be detected. A **receptor** is a specialised cell that can detect a particular stimulus. Large, complicated organisms such as humans have many different receptors for lots of different stimuli. With single-celled organisms, the whole organism is the receptor.

Now test yourself TESTED

1 Name five different stimuli that humans can detect.

Answer on p. 206

Tropisms

REVISED

Plants have no nerves or muscles, so they cannot respond as fast as animals. However, they are sensitive to the environment and can respond by growing in a particular direction. Growth responses to directional stimuli in plants are called **tropisms** and are achieved by changes in cell division and enlargement. In flowering plants, tropisms and other responses are controlled by specific **growth factors** — sometimes inaccurately called plant hormones. Generally, growth factors are made in the growing points — **meristems** — and move to other areas of the plant where they control mitosis and cell enlargement/specialisation.

A **tropism** is a growth response made by a plant in response to an external stimulus.

There are many tropisms, such as:

- **phototropism** — the response to light. Stems are usually positively phototropic and grow towards the light, whereas roots are generally negatively phototropic and grow away from the light.
- **geotropism** or **gravitropism** — both names for the response to gravity. Roots are usually positively geotropic and grown down into the soil, whereas stems are generally negatively geotropic and grow upwards. This allows seeds, which fall at random into soil, to grow in the right direction even in the absence of light.
- **hydrotropism** — the response to water
- **chemotropism** — the response to chemicals
- **thigmotropism** — the response to touch

Now test yourself TESTED

2 Explain the term *tropism*.

Answer on p. 206

The specific example of a growth factor you have to know is the control of phototropism and gravitropism by different concentrations of **indoleacetic acid (IAA)**. Figure 7.1 explains the phototropic response.

IAA belongs to a class of plant growth factors called **auxins**.

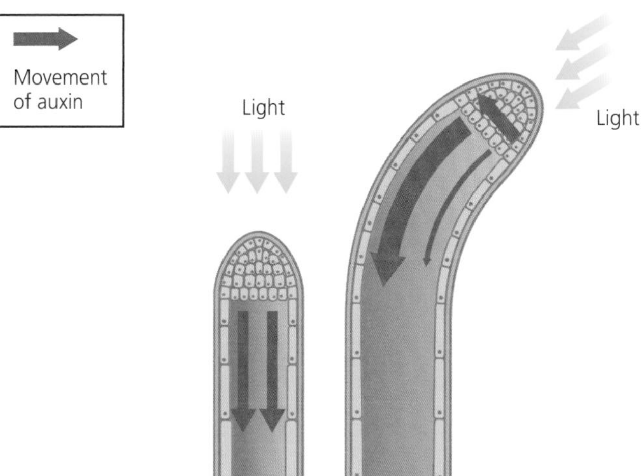

Figure 7.1 The mechanism of phototropism

1 Light illuminates the plant from one side.
2 The growing tip of the plant makes IAA, which is actively transported to the shaded side.
3 The IAA diffuses down the shaded side.
4 The IAA stimulates cell division and elongation.
5 The increased growth on the shaded side causes the stem to bend towards the light.

In gravitropism (Figure 7.2), the roots must grow downwards and the shoots must grow upwards. Imagine a root that is lying horizontally in the soil:

1 Cells near the root tip contain small organelles called amyloplasts, which are similar to chloroplasts but densely packed with starch.
2 Due to their density, the amyloplasts accumulate on the lower side of the cells and this allows the cells to detect the direction of gravity.
3 IAA seems to be actively transported to the area where the amyloplasts are. In roots, higher levels of IAA inhibit cell elongation (the opposite of the response in shoots).
4 As a consequence, the roots grow more quickly on the upper side, resulting in downward growth.

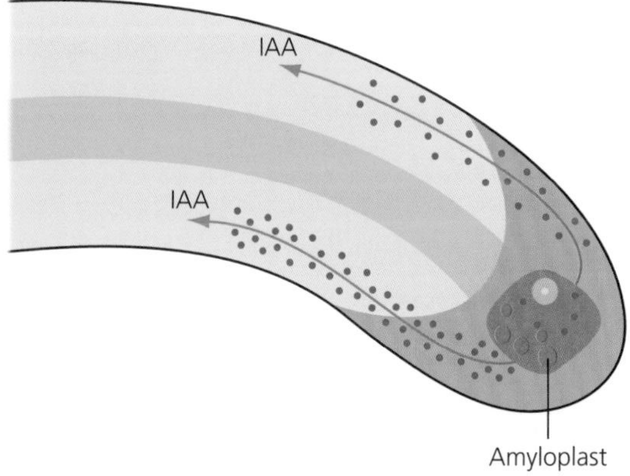

Figure 7.2 The mechanism of gravitropism in roots

Taxis and kinesis

REVISED

At their most basic level, responses consist of detecting particular stimuli and then moving or growing towards or away from them. For organisms that can move, such as bacteria, algae and most animals, **taxis** and **kinesis** are common responses.

A taxis is a **directional** response that results from an organism being able to tell the direction of a stimulus. If an organism moves towards a stimulus, it is a positive taxis. For example, some bacteria are positively **aerotactic** because they move towards oxygen.

In contrast, a kinesis is a **non-directional** response. This is an effective way of finding favourable conditions even when they cannot detect which direction to go. Woodlice, for example, prefer dark, damp conditions. When in light, dry conditions they move quickly, making many turns. In this way they explore their environment. When, by chance, they find the conditions they prefer, they slow down and turn more tightly, in an attempt to find the very best spot. Eventually, they stop.

The apparatus shown in Figure 7.3 is suitable for studying taxis and kinesis in small invertebrates such as woodlice and maggots.

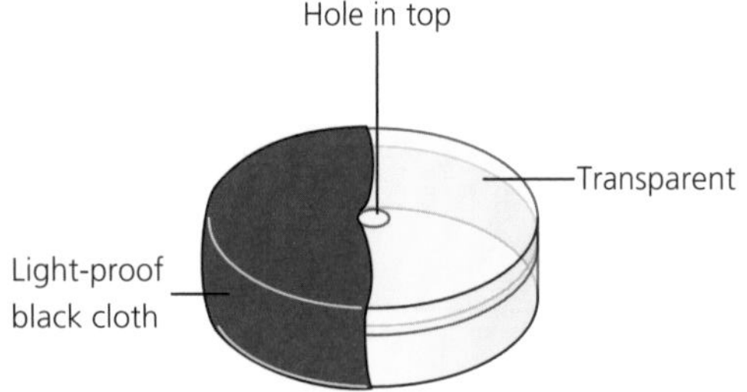

Figure 7.3 A simple choice chamber to investigate taxes or kineses in response to light

Now test yourself

TESTED

3. Explain the terms *taxis* and *kinesis*.
4. When observing an organism's behaviour in the choice chamber shown in Figure 7.3, explain how a taxis would differ from a kinesis.
5. Some single-celled algae have flagella so they can move, and are said to be positively phototactic.
   (a) Explain what *positively phototactic* means.
   (b) Explain the advantage of this characteristic to the algae.
6. Suggest the advantage to woodlice of seeking out damp, dark conditions.

Answers on p. 206

Reflex arcs

REVISED

Vertebrates have a complex nervous system that consists of a **central nervous system (CNS)** — the brain and spinal cord — along with many peripheral nerves. The basic organisation is as follows:

- **Receptors** are cells that gather information by detecting stimuli from within the body and the external environment.
- Sensory information passes into the CNS via impulses along **sensory nerves**.
- The CNS processes the information and **coordinates a response**.
- Impulses pass to effectors along **motor nerves**.
- The **effectors** — muscles or glands — bring about the response.

The simplest response we can study is the **reflex arc** (Figure 7.4), which involves three **neurones**:

- a **sensory neurone**
- a **relay neurone** inside the spinal cord
- a **motor neurone** (see page 138)

Neurones are nerve cells that transmit an action potential.

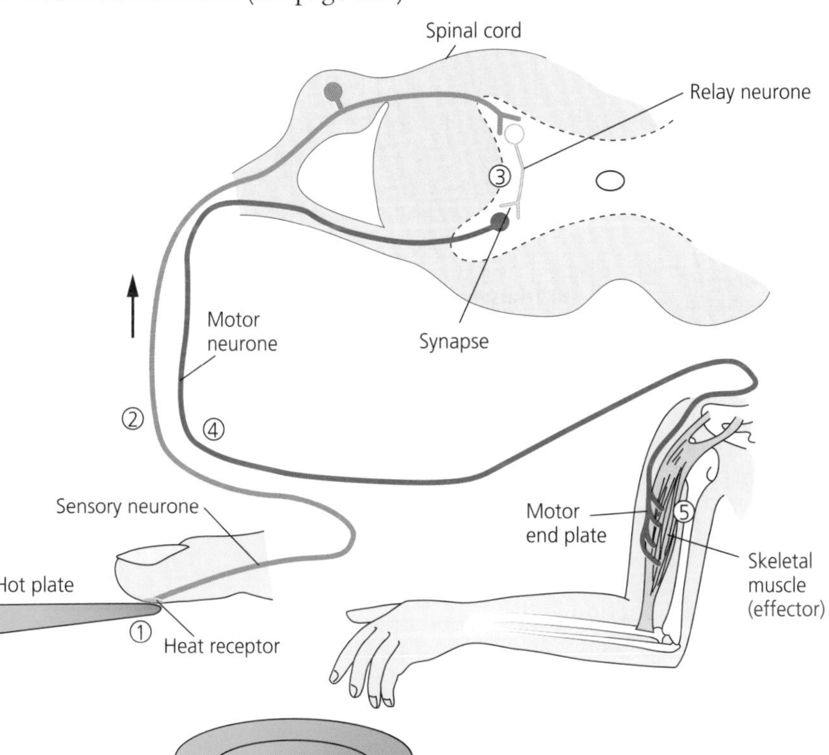

Figure 7.4 A reflex arc, showing the three basic neurones

Exam tip

In a reflex arc, you do *not* need to know about dorsal and ventral roots and ganglions. Lots of diagrams contain them.

Reflex arcs are simple for good reason: it makes them as fast as possible. Generally, reflexes are fixed, which means that a particular stimulus always produces the same response. This is because they do not involve the conscious parts of the brain. The two main functions of reflex arcs are to minimise damage to the body and to maintain posture by making constant minute adjustments to muscle tone so that we remain balanced. An example of a three-neurone reflex is the knee-jerk reflex, which is important in helping to maintain posture.

Revision activity

Draw a flow diagram of the withdrawal-from-heat reflex arc. Start with 'finger touches hot plate' and end with 'hand is snatched away'.

Receptors

The Pacinian corpuscle

REVISED

The **Pacinian corpuscle** is one example of the different receptors found in mammalian skin. Its function is to detect heavy pressure and vibration (Figure 7.5).

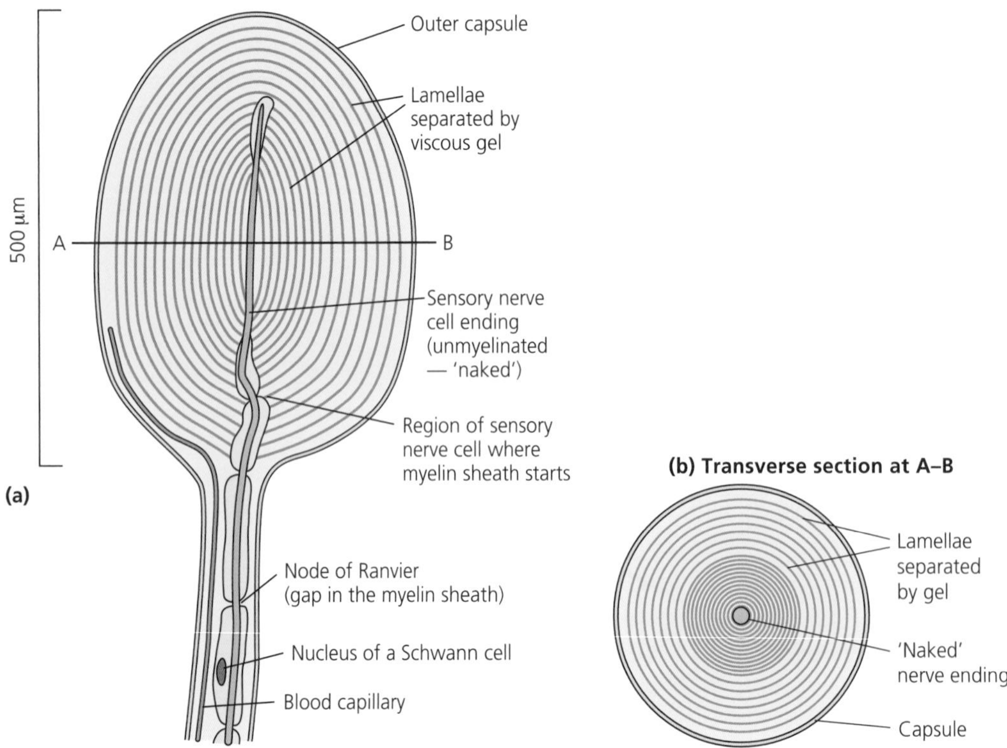

Figure 7.5 A Pacinian corpuscle in (a) longitudinal section and (b) transverse section

Each corpuscle is about 1 mm in diameter and consists of 20–60 lamellae (layers of cell membrane) each separated by a gel, so its structure resembles an onion. At the centre is a single receptor cell, often called a nerve ending. In the membrane of the receptor are proteins called **stretch-mediated sodium channels**. The corpuscle works as follows:

1 Pressure causes the proteins to change shape and become more permeable to sodium ions.
2 Sodium ions rapidly diffuse in, creating a generator potential.
3 If the generator potential reaches a threshold value, an **action potential** is generated in the sensory nerve and we get the sensation of pressure.
4 The receptor adapts very quickly, so it does not generate prolonged impulses. The gel quickly redistributes to even out the pressure, so the membrane proteins return to their normal permeability.

The vital points about this receptor (and most receptors) are:

- It only responds to one type of stimulus.
- Stronger stimuli generate more frequent impulses.
- It stops responding to a prolonged stimulus.

Now test yourself

TESTED

7 Suggest the advantage of receptors adapting, so that they do not respond to prolonged stimuli.

Answer on p. 206

The retina

REVISED

The retina is the layer of light-sensitive receptor cells — **rods** and **cones** — that gather visual information and channel it down the **optic nerve** to the brain, where an image is formed. There are two key aspects to the retina:

- **sensitivity** — the ability to see in dim light
- **visual acuity** — the ability to see detail

The rods are more sensitive because several of them **converge** into one sensory neurone, allowing them to **summate** (Figure 7.6). All the rods supplying the neurone can contribute to reaching the **threshold value**. The cones have a higher visual acuity because there is much less convergence — at the centre of the **fovea** each cone has its own sensory neurone. Consequently, the cones send more information to the brain per unit area of retina.

The **fovea** is part of the retina that has no rod cells but large numbers of cone cells. Because it has a large number of cone cells, it is the region of highest visual acuity. In other words, it is the part of the retina that enables the greatest degree of detail to be seen.

Exam tip

Visual acuity can be thought of as resolution — the ability to distinguish two close objects. Remember that resolution was a key factor in the power of a microscope.

Typical mistake

Thinking that cones must be more sensitive because they can perceive more detail. Rods are more sensitive because they can function in lower light intensities.

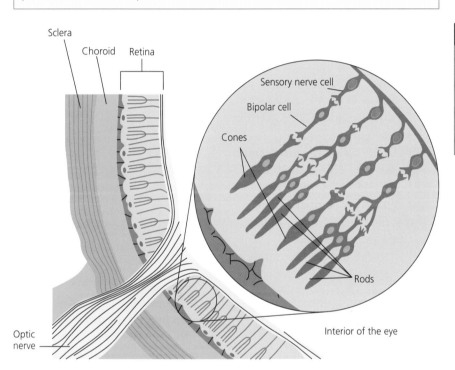

Figure 7.6 The distribution of rods and cones in the retina. Note that several rods (shown here as three) converge into one sensory nerve cell, whereas the cones do not converge

Table 7.1 Comparing rods and cones

Feature	Rods	Cones
Distribution	Periphery of retina	Concentrated at fovea; a few in the periphery
Sensitive to	Dim light	Bright light
Amount of retinal convergence	Many rods converge into one neurone	Each cone has its own neurone
Overall function	Vision in poor light	Vision in colour and detail in good light

Now test yourself

TESTED

8 Explain what is meant by the term *visual acuity*.

Answer on p. 206

Control of heart rate

The heart is **myogenic**, which means that the impulse that causes the beat originates within the heart muscle itself. Individual heart muscle cells beat on their own. The control of heartbeat has two aspects:

- Making sure that the heart muscle cells all beat at the right time, in the following way: both atria together → delay → both ventricles together.
- Speeding up or slowing down the heart rate according to the needs of the body.

The conducting pathway of the heart

REVISED

All the heart muscle cells need to contract at the right time for the heart to work. Chambers should only contract when they are full of blood, so the heart has a conducting pathway of specialised muscle fibres to ensure the right sequence of events. The atria must contract first and then, when full, the ventricles follow. This means a delay is needed to allow the ventricles to fill. The full sequence is as follows (Figure 7.7):

1. The heartbeat is initiated by a group of cells called the **sinoatrial node (SAN)** near the top of the right atrium. This is the body's pacemaker.
2. These cells produce waves of **electrical activity**, similar to nerve impulses.
3. The impulse spreads over the atria, which then contract.
4. A tough band of connective tissue prevents the impulse spreading to the ventricles.
5. The impulse is picked up by the **atrioventricular node (AVN)** which, after a short delay, passes the impulse down the middle of the ventricles in the **bundle of His** — a specialised bunch of muscle fibres that transmits the impulse without causing contraction.
6. At the apex of the heart (the bottom of the ventricles) the impulse reaches the **Purkinje fibres**. This causes contraction of the thick ventricle muscle, starting at the apex, so that blood is forced upwards through the semilunar valves.

Exam tip

An important difference: the bundle of His transfers the impulse without causing contraction, whereas the Purkinje fibres cause contraction.

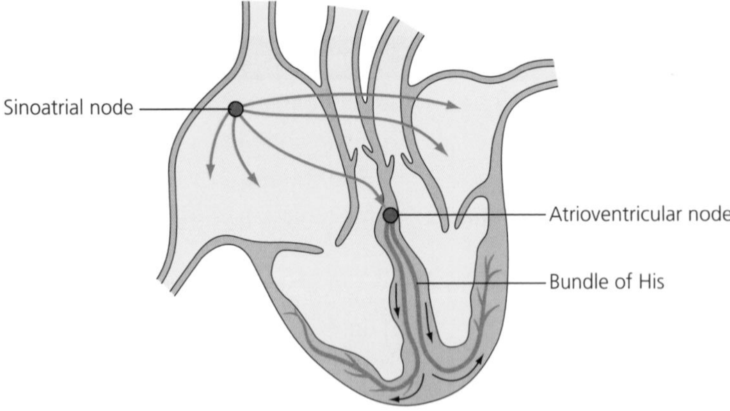

Figure 7.7 The route of electrical activity that makes the heart beat in a smooth sequence

Now test yourself

TESTED

9 Explain what is meant by the term *myogenic*.
10 Explain why it is important that there is a slight delay after the atria contract.

Answers on p. 206

The role of the autonomic nervous system in the control of heart rate

REVISED

The **autonomic nervous system (ANS)** consists of two sets of nerves that control many homeostatic functions without involving conscious thought. Autonomic means 'self-governing'. The two sets of nerves are called **sympathetic** and **parasympathetic**.

Generally, impulses down sympathetic nerves will prepare the body for action, in much the same way as adrenaline. The parasympathetic nerves act **antagonistically** to the sympathetic nerves, bringing the body back to normal and generally having a calming effect. Sympathetic nerves release a **neurotransmitter** called **noradrenaline** from their **synapses**, which is similar to the hormone adrenaline. Parasympathetic synapses release **acetylcholine (ACh)** from their synapses.

A **neurotransmitter** is a substance that transmits information across the synaptic cleft between two neurones.

A **synapse** is a junction between nerve cells.

Acetylcholine (ACh) is a common neurotransmitter responsible for the transmission of a nerve impulse across a synapse. There are many other neurotransmitters, especially in the brain.

Exam tip

The responses brought about by the sympathetic nervous system are often described as 'fight or flight', while the responses brought about by the parasympathetic nervous system are described as 'rest and digest'. These are not expressions to be used in exams.

Chemoreceptors and baroreceptors

REVISED

A good example of the ANS in action is the control of heart rate. The rate of heartbeat is controlled by an area of the brain called the **cardioregulatory centre**, which is located in the **medulla oblongata** in the hindbrain (Figure 7.8). Within the cardioregulatory centre there is an **acceleratory centre** and a separate **inhibitory centre**. In order to match the heart rate to the demands of the body, the cardioregulatory centre gathers information from two types of receptor (Figure 7.9):

- **chemoreceptors** — chemical receptors that detect carbon dioxide concentration in the blood. These receptors are found in two places: the **carotid bodies** in the walls of the carotid arteries in the neck and the **aortic body** in the aorta, just above the heart.
- **pressure receptors** — also called baroreceptors — in the wall of the carotid sinus, a small swelling in the carotid artery. The higher the blood pressure, the more the baroreceptors send impulses to the cardioregulatory centre. This mechanism prevents blood pressure from going too high. It can also speed up the heart if it detects that blood pressure is going too low.

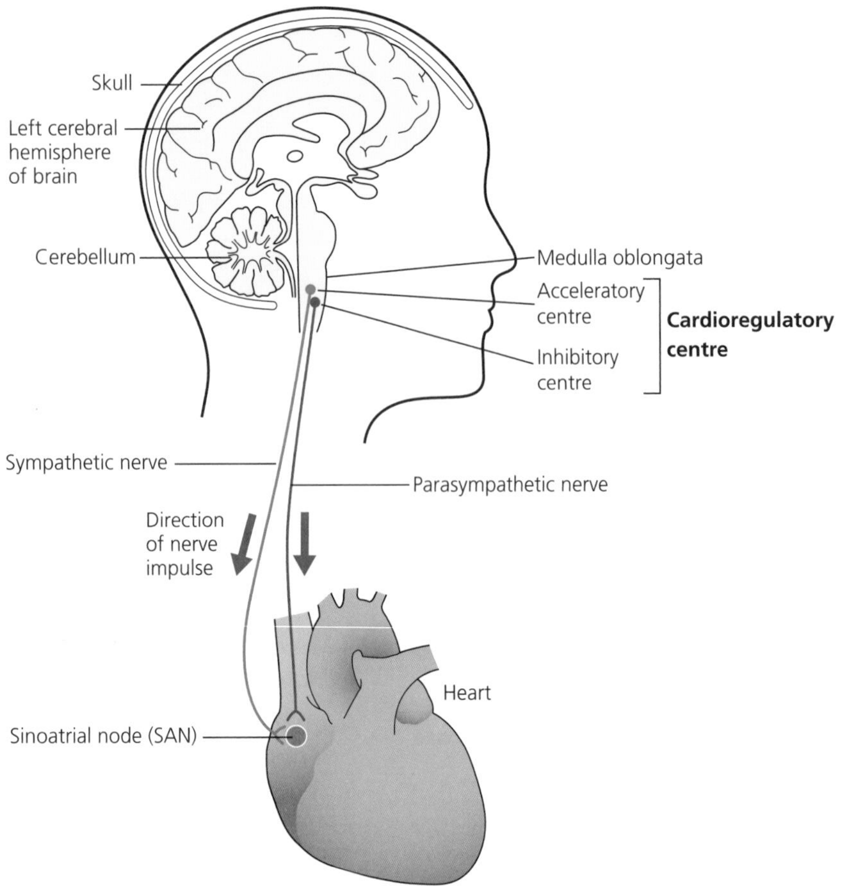

Figure 7.8 The cardioregulatory centre and the nerve supply to the heart

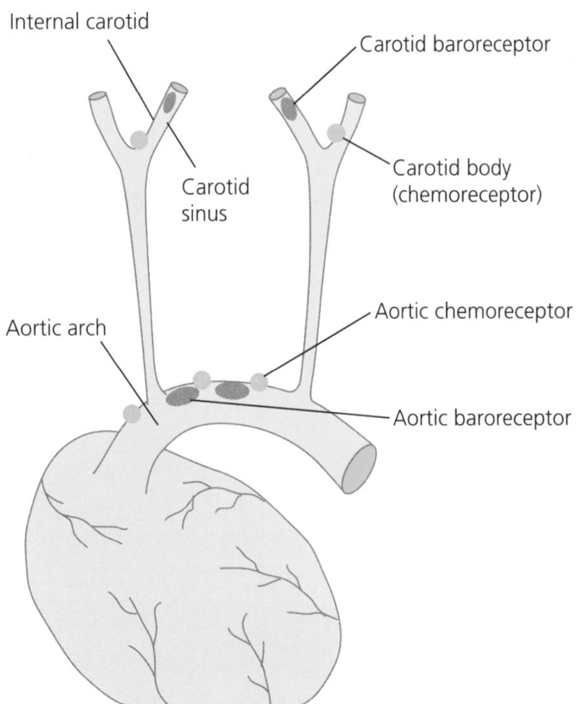

Figure 7.9 The location of chemoreceptors and baroreceptors in blood vessels above the heart

Nerve supply to the heart

REVISED

There are two nerves that pass from the cardioregulatory centre to the **sinoatrial node (SAN)** in the heart:

- the sympathetic or accelerator nerve
- the parasympathetic or decelerator nerve

The SAN sets a regular beat, but it is modified by these two nerves. When the chemoreceptors detect an increase in carbon dioxide levels, impulses pass to the **cardioregulatory centre**, which responds by sending impulses down the sympathetic nerve. The synapse releases noradrenaline onto the SAN, which speeds up the frequency of the impulses it generates. The impulses are picked up by the atrioventricular node (AVN), which creates a delay, allowing the ventricles to fill, before channelling impulses down the bundle of His and into the Purkinje fibres. This causes ventricular contraction.

In practice, the control of heart rate is more complex and fine-tuned as a result of the balance of impulses down both nerves. There is also an element of anticipation so that exercise stimulates an increase in heart rate and breathing before an increase in carbon dioxide is detected.

Revision activity

Draw a flow diagram to summarise how heart rate slows down after exercise.

Exam practice

1 (a) Define the term 'stimulus'. [2]

Pacinian corpuscles are receptors found in the skin.

(b) Explain how pressure applied to a Pacinian corpuscle can lead to the generation of a nerve impulse. [3]

(c) If a constant stimulus is applied to the receptor, it stops responding. Suggest the advantage of this. [1]

2 An investigation was carried out to see how a particular species of blowfly larva ('maggot') responded to light. The investigator drew the diagram below and placed it on a piece of paper. The paper was placed in a dark room with just a single light source. The investigation was carried out several times, using a different maggot and a new piece of paper each time. The segment the maggot was in after 30 seconds was recorded.

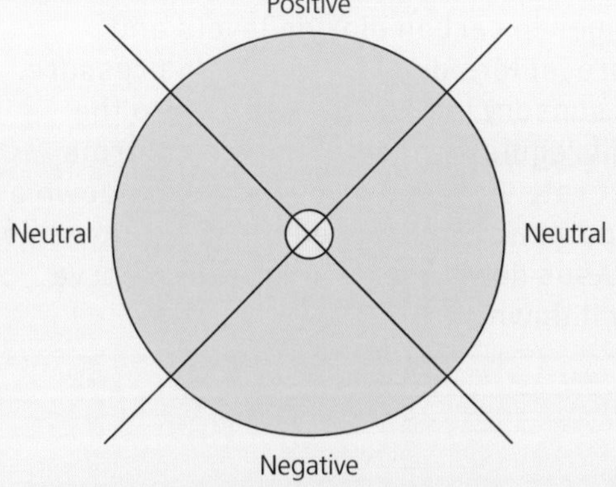

(a) Show on the diagram where the light source should be placed. [1]

(b) Describe the difference between a taxis and a kinesis. [2]

(c) Suggest why a new piece of paper was used each time. [2]

(d) The investigator decided that if the maggots were not responding to light, there should be approximately equal numbers of maggots ending up in the positive, negative and neutral segments.

(i) Suggest a suitable statistical test for this investigation. Explain your choice. [2]

(ii) The test statistic was calculated and found to be significant at the 0.01 level. Explain what this means. [2]

3 The following graph shows the distribution of rod and cone cells across the retina of the human eye.

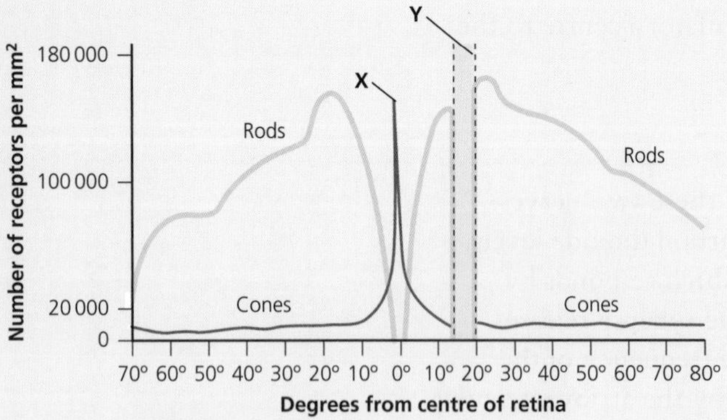

(a) Explain why cones have a greater visual acuity than rods. [2]

(b) We can only see in detail, such as the exact words you are reading on this page, because an image of each word falls on point X. Explain why point X allows such a high visual acuity. [2]

(c) Explain why, when an image is formed on area Y of the retina, nothing is perceived by the brain. [2]

(d) When looking at a particularly dim star, it appears clearer if we do not look straight at it. Use your knowledge of rod cells to explain why. [3]

Answers and quick quiz 7 online

ONLINE

Summary

By the end of this chapter you should be able to understand the following:

- Tropisms are simple growth responses in plants, controlled by indoleacetic acid (IAA).
- Taxes and kineses are simple responses that allow mobile organisms to find favourable conditions. Taxes are directional, kineses are not.
- A simple reflex arc is the simplest coordinated response and involves three neurones. Reflexes generally exist to avoid damage or to maintain posture.
- The Pacinian corpuscle is an example of a receptor. It is found in the skin and detects pressure and vibration.
- In the retina, the differences in sensitivity and visual acuity can explained by the distribution of rods and cones and the connections they make in the optic nerve. Rods have greater sensitivity but cones have greater acuity.
- The conducting pathway of the heart ensures that both atria contract at the same time and then, after a delay, ensures that both ventricles contract at the same time.
- Heart rate is controlled by the autonomic nervous system. Chemoreceptors detect changes in carbon dioxide levels and baroreceptors detect changes in pressure. This sensory information is fed into the cardioregulatory centre, which controls heart rate via two nerves. Impulses down a sympathetic nerve speed the heart up while impulses down a parasympathetic nerve slow it down.

8 Nervous coordination

Nerve impulses

The structure of a motor neurone

REVISED

Neurones are specialised cells that transmit impulses. Motor neurones (Figure 8.1) transmit impulses from the central nervous system to an effector, such as a muscle or a gland. Their key features are:

- a cell body that contains the nucleus and other organelles
- an elongated axon that carries impulses away from the cell body
- one or more dendrites that take impulses towards the cell body
- a myelin sheath

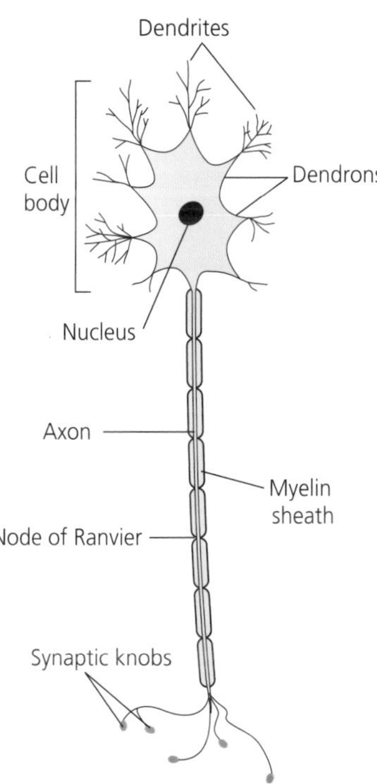

Figure 8.1 The basic structure of a motor neurone

Now test yourself

TESTED

1 Where in the neurone is the nucleus?
2 What is the difference between an axon and a dendrite?

Answers on p. 206

Many neurones are **myelinated**, which means that their axon is protected by a fatty **myelin sheath** made from several layers of cell membrane. The sections are made by individual Schwann cells that wrap around the axon many times. In between the sections of myelin are nodes of Ranvier, where the axon membrane is exposed. The role of the myelin sheath is to:

- insulate the axon
- physically protect the axon
- speed up the transmission of nerve impulses (page 140–3)

A **myelin sheath** is a non-conducting fatty layer around the neurone.

The resting potential

REVISED

When a nerve impulse is not being conducted, an **electrochemical gradient** is maintained between the inside and the outside of the axon membrane. This difference in charge is caused by different numbers of positive and negative ions. It is known as the **resting potential** and is a state of readiness in a neurone. In a resting cell, there is a potential difference across the membrane of about −70 mV. Once the resting potential is established, the neurone is ready to transmit an impulse. The resting potential results from two processes occurring together: active transport and unequal facilitated diffusion:

- **Active transport** — the sodium/potassium pump is a protein responsible for the active transport of positive ions. For each ATP molecule split, three **sodium** (Na^+) ions are pumped out of the axon and two **potassium** (K^+) ions are pumped in (Figure 8.2).
- **Unequal facilitated diffusion** — there are sodium channels and potassium channels. Both are gated so that, by changing their shape, they can open up and allow the ions to diffuse freely. The key difference is that, at rest, the potassium channels are more 'leaky' than the sodium channels, so the potassium diffuses out faster than sodium diffuses in.

An **electrochemical gradient** is simply a difference in charge on two sides of a membrane. It is caused by different concentrations of positive and negative ions.

The **resting potential** is the potential difference in electrical charge across the cell surface membrane of a nerve cell when the cell is at rest.

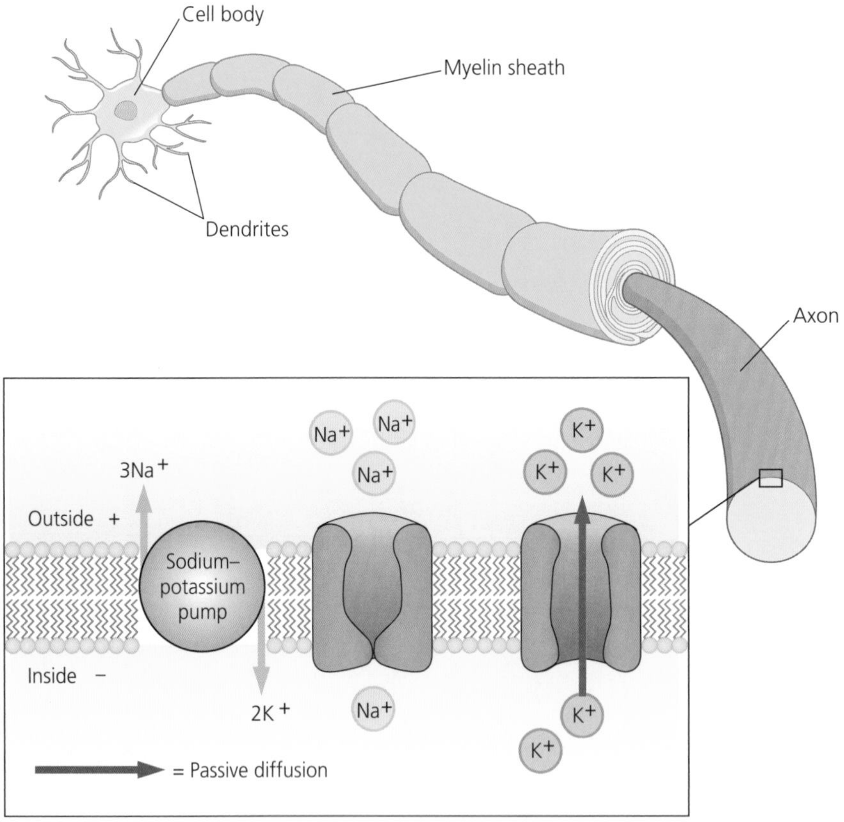

Cell body
Myelin sheath
Dendrites
Axon
3Na+
Outside +
Sodium–potassium pump
Inside −
2K+
Na+
K+
= Passive diffusion

Figure 8.2 The resting potential across the axon membrane

The overall result is a high concentration of positive ions outside the axon, giving the area a positive charge relative to the inside. This is the resting potential. There are also negative ions inside and outside the axon membrane, the most common being chloride ions and negatively charged proteins. However, it is the movement of positive ions that establishes the resting potential and brings about the **action potential**, so we focus on those.

Revision activity

Look at an animation of the mechanism of resting potential and action potential — there are lots available.

Now test yourself

TESTED

3 What are the two key processes that result in the creation of the resting potential?
4 What is the value of the average resting potential in mV (millivolts)?

Answers on p. 206

The action potential

REVISED

The action potential is the **nerve impulse**. It is a rapid reversal of the resting potential that spreads rapidly along the axon. It is started by one simple action: the gated sodium ion channels open for a fraction of a second (Figure 8.3). For about a millisecond, the resting potential is reversed in one area of the axon so that the inside becomes positively charged (with respect to the outside). This reversal spreads quickly along the axon while the original area recovers and establishes the resting potential again.

Typical mistake

Nerve impulses are not messages. 'Nerve impulse' and 'action potential' are good A-level terms, but don't call them messages or signals.

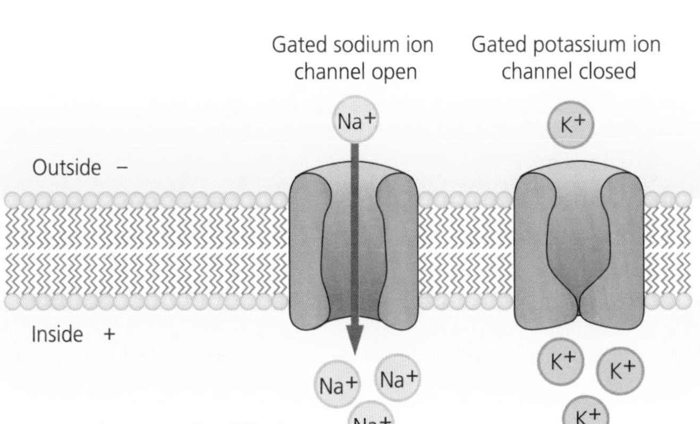

Figure 8.3 The action potential is initiated by the opening of gated sodium ion channels

The key events are as follows:

1 Depolarisation — the gated sodium ion channels open, allowing sodium ions to diffuse in and reverse the resting potential.
2 Repolaristion — the gated sodium ion channels close, preventing further diffusion inwards. The gated potassium ion channels open fully, allowing positively charged potassium ions to diffuse out rapidly.
3 Hyperpolarisation — an 'overshoot' results from the gated potassium ion channels still being fully open while the active transport mechanism begins to re-establish the resting potential by pumping sodium ions out.
4 The gated potassium ion channels return to their normal permeability and the resting potential is re-established.

These events, which occur over a short period of time at one place in the axon, are shown in Figure 8.4.

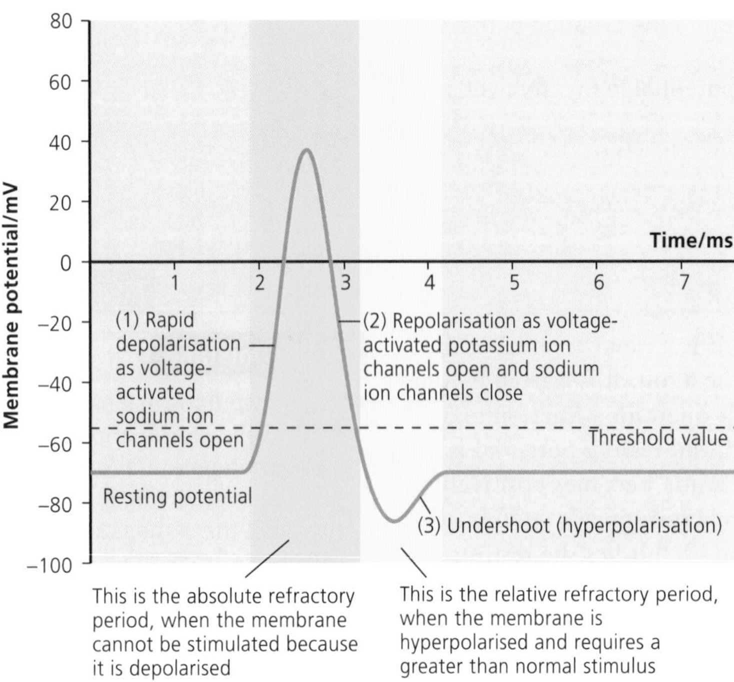

Figure 8.4 An action potential trace

The action potential is **self-propagating**, which means that the depolarisation of one region of an axon will immediately cause the depolarisation of the next region. The gated sodium and potassium ion channels in the axon membrane are described as 'voltage-gated', which means that their shape, and therefore their permeability, depends on the voltage (charge) across the membrane.

The all-or-nothing principle

REVISED

If the generator potential is large enough, and its **threshold value** is reached, an action potential is generated. There either *is* an action potential or there *is not*. If the generator potential is not large enough, there is no action potential. This is the **all-or-nothing** principle. The brain makes sense of the incoming sensory information. It knows:

- the origin of the incoming impulses — if the stretch receptors in your bladder send impulses to the brain, you know what it means and what you need to do
- the frequency of the incoming impulse — the more intense the stimulus, the more frequent the impulses. It is not possible to have a small action potential for a small stimulus and a larger one for a large stimulus.

All-or-nothing describes a nerve impulse. When an action potential is produced in a nerve cell, it is always the same size. It does not matter how big the initial stimulus, the action potential will always involve the same change in potential difference across the cell surface membrane.

The refractory period

REVISED

After an impulse has passed, there is a short period of time when it is impossible to initiate a new action potential. This is the **absolute refractory period**. This is followed by a brief period when it is possible to generate an impulse but the threshold value is greater, so the stimulus must be of greater intensity than normal. This is the **relative refractory period**. Both phases of the refractory period are shown in Figure 8.4.

The **relative refractory period** is the short recovery period that occurs immediately after the passage of a nerve impulse along the axon of a nerve cell.

The refractory period is important because it keeps impulses discrete. Without it, they would tend to blend into each other and produce continuous stimulation. The refractory period limits the frequency of impulses that can be transported.

Factors affecting the speed of conductance

REVISED

Several factors affect the speed of conductance of a nerve impulse:

- The conduction of impulses along myelinated neurones is faster than in non-myelinated neurones because the impulse jumps from one node to the next. This is called **saltatory conduction**. Myelin prevents the depolarisation from happening, so it only happens at the nodes where the axon membrane is exposed.
- The wider the **axon diameter**, the faster the transmission. Generally, this is only important in invertebrates, which have never evolved myelin.
- The higher the **temperature**, the faster the transmission because the movement of ions is faster.

The speed of transmission from one part of the body to another is also affected by the number of synapses in any given pathway. Synapses slow transmission down, which is why there are as few as possible in a reflex arc.

Synaptic transmission

A **synapse** is a junction between two neurones or between a neurone and a muscle, in which case it is called a **neuromuscular junction**. Synapses are found at the ends of axons (Figure 8.5).

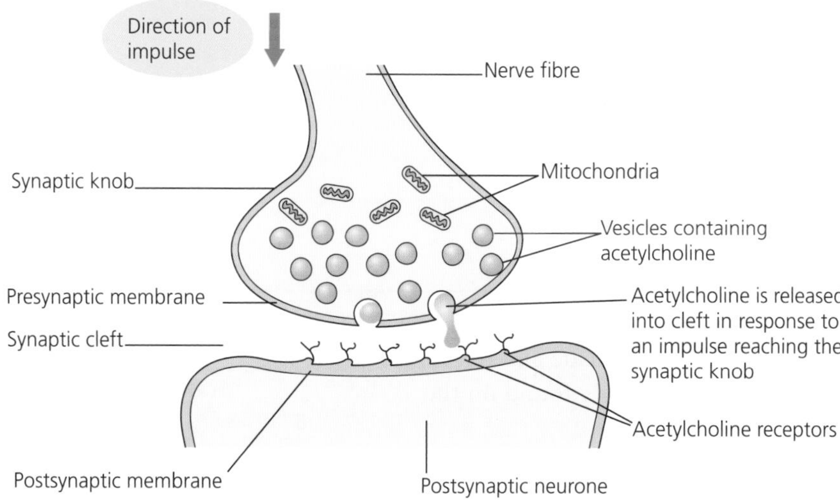

Figure 8.5 The structure of a synapse

The events of synaptic **transmission** are shown in Figure 8.6. Immediately after transmission, an enzyme in the **synaptic cleft** breaks down the neurotransmitter. The products of this breakdown are reabsorbed into the presynaptic membrane to be resynthesised into the active transmitter again. In many synapses outside of the central nervous system, the neurotransmitter is **acetylcholine (ACh)**. These synapses are said to be **cholinergic synapses**. There are many other transmitter substances, especially in the brain.

The **synaptic cleft** is the small gap between the presynaptic neurone and the postsynaptic neurone at a synapse.

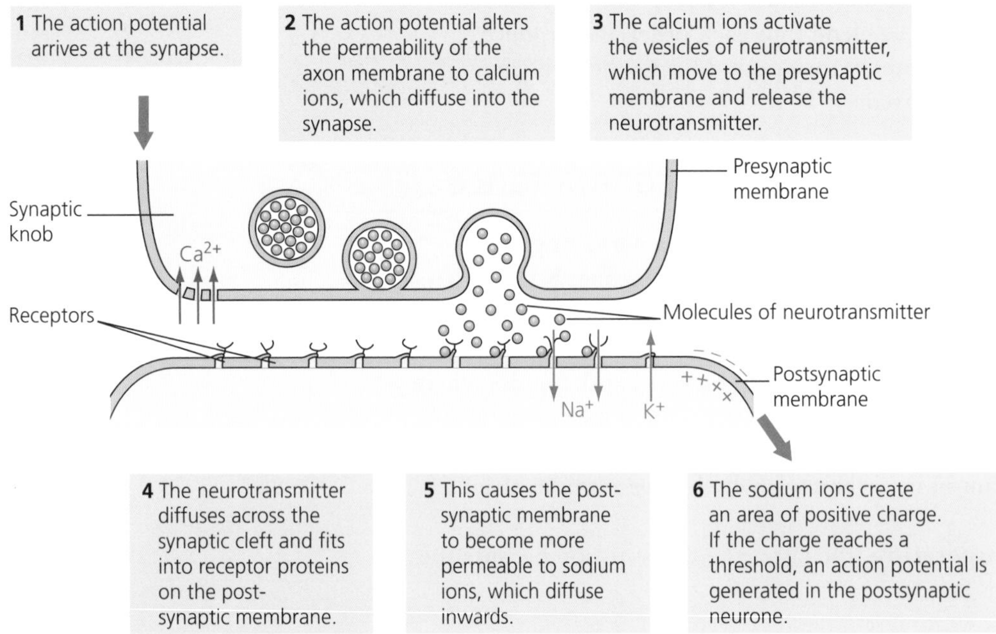

1 The action potential arrives at the synapse.
2 The action potential alters the permeability of the axon membrane to calcium ions, which diffuse into the synapse.
3 The calcium ions activate the vesicles of neurotransmitter, which move to the presynaptic membrane and release the neurotransmitter.
Presynaptic membrane
Synaptic knob
Ca2+
Receptors
Molecules of neurotransmitter
Postsynaptic membrane
Na+
K+
4 The neurotransmitter diffuses across the synaptic cleft and fits into receptor proteins on the post-synaptic membrane.
5 This causes the post-synaptic membrane to become more permeable to sodium ions, which diffuse inwards.
6 The sodium ions create an area of positive charge. If the charge reaches a threshold, an action potential is generated in the postsynaptic neurone.

Figure 8.6 Synaptic transmission

Now test yourself

TESTED

5 Fill in the gaps in the following flow passage to describe synaptic transmission.

The action potential arrives at the synapse → _______ ions flow into the presynaptic membrane → Molecules of neurotransmitter _______ across the synaptic cleft and fit into specific _______ → The _______ of the postsynaptic membrane changes → _______ ions flow in, causing a positive charge to build up inside the _______ synaptic membrane → If the _______ is reached, an action potential is created in the postsynaptic neurone.

6 Suggest *two* reasons for the presence of mitochondria in the synaptic knob.

Answers on p. 206

Key features of synapses

REVISED

Synapses are **unidirectional**, which means an impulse can only pass one way. This is because the neurotransmitter can only be made and released on the presynaptic side and the receptor proteins are only found on the postsynaptic side.

Some synapses are **inhibitory**. It is just as important to be able to switch synapses off, so that precise pathways can be selected. An inhibitory synapse makes it less likely that an impulse is generated in the postsynaptic membrane. As shown in Figure 8.7, neurones A and B are stimulatory synapses but neurone C is an inhibitory synapse.

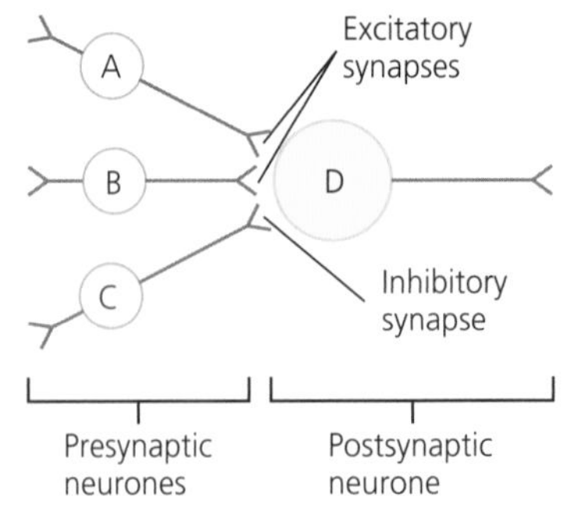

Figure 8.7 The effect on the postsynaptic neurone

Synapses can also **summate**, which means they combine to generate an impulse. There are two types of summation:

- **temporal summation** (separated in time) — impulses are transmitted down the same synapse in quick succession
- **spatial summation** (separated in space) — impulses are transmitted down two or more separate synapses at the same time

Table 8.1 Summation at a synapse (see Figure 8.7)

Impulses down presynaptic neurones	Impulse generated in D?	Explanation
One impulse down A	No	Threshold is not reached
One impulse down B	No	Threshold is not reached
Two impulses down A in quick succession	Yes	Temporal summation — both impulses combine to reach the threshold
Two impulses down A and B at same time	Yes	Spatial summation — both impulses combine to reach the threshold
Three impulses down A, B and C at same time	No	The threshold is not reached because neurone C is inhibitory, so it cancels out one excitatory impulse

Now test yourself TESTED

7 Suggest why it is important that synapses are unidirectional.

Answer on p. 206

Comparison of a cholinergic synapse and a neuromuscular junction

REVISED

The major structural difference between the two is that a synapse connects one neurone to another neurone, while a neuromuscular junction connects a neurone to a bundle of muscle fibres (known as a motor unit). The basic mode of action is the same in the presynaptic part, but the key difference is in the postsynaptic part. In a synapse, there is a build-up of a postsynaptic potential, which can lead to the generation of another nerve impulse. In a neuromuscular junction, there is a wave of depolarisation that spreads along the muscle fibres, setting in motion the sequence of events that leads to muscular contraction.

The effect of drugs

REVISED

Many **drugs** (legal or otherwise) exert their effect by interfering with normal synaptic transmission. For example, some drugs:

- block the calcium channels
- prevent synthesis of the neurotransmitter
- fit into the postsynaptic receptors
- prevent re-uptake of the neurotransmitter
- block the enzyme that breaks down the neurotransmitter

Exam tip

You don't need to learn the action of any specific drugs. Exam questions will test your understanding of synapses by telling you how specific (and probably unfamiliar) drugs work.

Now test yourself TESTED

8 For each of the effects of drugs listed in the bullet list above, predict the effect on synaptic transmission.

Answer on p. 206

Skeletal muscles

Types and function of skeletal muscle

REVISED

Skeletal muscles produce movement and maintain posture. They are attached to bones by tendons and are generally under conscious control. Muscle is a specialised tissue that can do just one thing — contract — and so skeletal muscles generally work antagonistically in pairs or groups. One contracts while the other relaxes.

There are two other types of muscle: **smooth muscle**, which is found in tubular organs such as the gut, reproductive system and blood vessels, and **cardiac muscle**, which is found only in the heart.

The sliding filament theory of muscle contraction

REVISED

Skeletal muscle is made from many small cylindrical fibres called **myofibrils**. Along their length is an alternating pattern of light and dark bands caused by two key interlocking proteins, **actin** and **myosin**. Each repeated unit of actin and myosin is called a **sarcomere**. Muscular contraction is caused when myosin pulls itself into the actin, shortening all the sarcomeres and therefore the muscle as a whole.

Myofibrils are the small fibres that are arranged parallel to each other in a skeletal muscle fibre.

In a resting muscle, a thin fibrous protein called **tropomyosin** covers the sites where actin binds to myosin. **Calcium ions** are essential for contraction, but they are stored outside the myofibril, in the **sarcoplasmic reticulum**. Figure 8.8 explains the theory.

(a) A section through a muscle fibre showing the myofibrils. The sarcoplasmic reticulum (in blue) stores the calcium ions essential in contraction

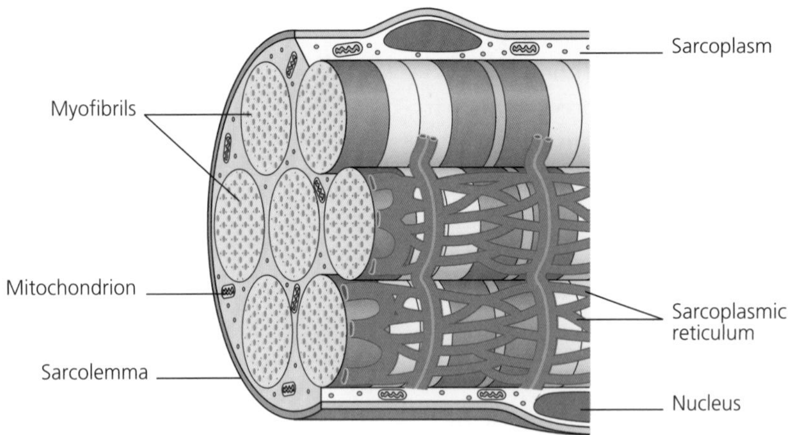

(b) The repeated pattern of actin and myosin fibres in two sarcomeres

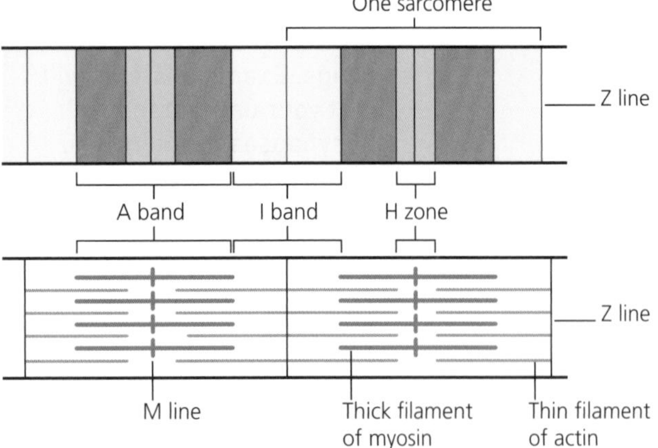

(c) The same sarcomeres after contraction

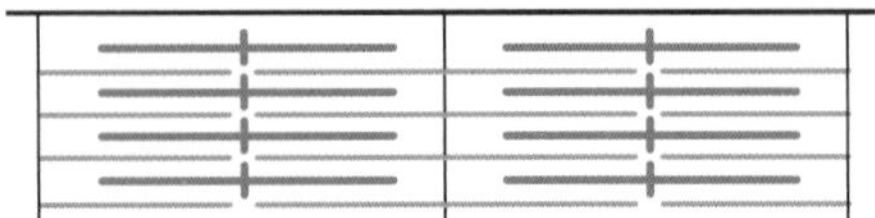

Figure 8.8 The sliding filament theory of muscle contraction

Exam tip

Don't worry about learning the names of the different bands — I band, A band etc. Exam questions will ask you about the mechanism of contraction and how the sarcomere will change.

How does a muscle contract?

When a muscle fibre is stimulated by nerve impulses, the following steps occur resulting in contraction of the muscle (Figures 8.9 and 8.10):

1 An **action potential** arrives down a motor neurone, reaching the **neuromuscular junction**.
2 In a similar mechanism to synapses, the neurotransmitter **acetylcholine (ACh)** is released onto the motor endplate.
3 The neurotransmitter causes a wave of **depolarisation** to spread along and around the myofibrils in the sarcoplasmic reticulum and T tubules.
4 Calcium ions are released from the sarcoplasmic reticulum and into the myofibril.
5 The calcium ions bind to **troponin** molecules, a small globular protein that binds to **tropomyosin** and moves it away from the actin–myosin binding site.
6 The **myosin heads** can bind to the actin molecules and form **actinomyosin bridges**.
7 The **ATP** attached to the myosin head splits. This releases the energy that makes the myosin head bend, pulling the actin along.
8 Another ATP molecule attaches to the myosin head and splits. The energy is used to detach the myosin head, change the angle of the head ('re-cock') and re-attach it further along the actin.
9 The process repeats and the myosin pulls itself over the actin, shortening the sarcomere.

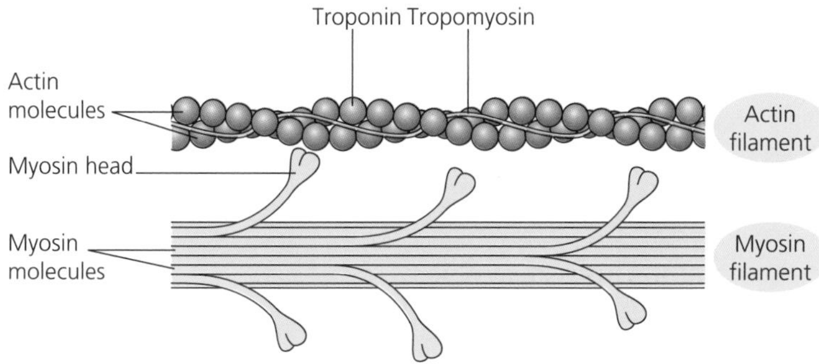

Figure 8.9 The arrangement of the four major proteins in muscular contraction

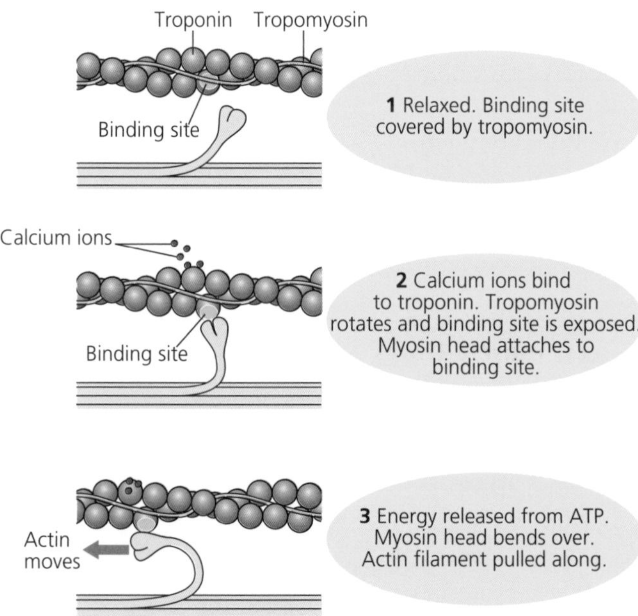

Figure 8.10 The key stages of contraction

Now test yourself

TESTED

9 Complete the following table by summarising the roles of the different components in muscular contraction.

Component	Role in contraction
Actin	
Myosin	
Troponin	
Tropomyosin	
ATP	
Calcium	

Answer on p. 206

ATP and phosphocreatine

REVISED

Muscles cannot move without ATP, but at any one time there is not much available and it doesn't last long. Prolonged exercise needs a continuous supply.

ATP can come from three sources, depending on the timescale:

- Initially, we use the ATP already present in the muscles, which has accumulated during periods of rest. However, it only lasts for a few seconds during vigorous exercise. There is another chemical, **phosphocreatine (PC)**, which can be hydrolysed instantly to resynthesise ATP. This ATP/PC system provides ATP for about the first 10 seconds of vigorous exercise.
- ATP from **glycolysis** — the first stage of respiration. This is anaerobic respiration, which yields only 2 ATP molecules per glucose molecule and comes with the problem of lactate build-up. However, it is a relatively quick process and bridges the gap between the first and last sources, i.e. between 10 seconds and about 1 minute.
- **Full aerobic respiration** — the reactions that take place in the mitochondria can provide an extra 34 ATP molecules per glucose molecule and there is no lactate build-up. However, it is slow and can only provide ATP at a certain rate. This is why you cannot sprint for long distances. The first two sources can provide ATP for contraction at full power, but the aerobic system can only provide ATP for about 60–70% of the maximum.

Fast and slow skeletal muscle fibres

REVISED

There are two types of skeletal muscle fibre, **fast skeletal muscle fibres** and **slow skeletal muscle fibres**, named according to their speed of contraction. The balance of the two types in our muscles is genetically determined, although training can alter the balance. Some are born with a lot of slow or fast muscle fibres, which makes them natural athletes at either endurance or strength events.

Table 8.2 Comparison of slow and fast skeletal muscle fibres

Feature	Fast skeletal muscle fibres	Slow skeletal muscle fibres
Speed of contraction	Fast, powerful	Slow
Speed of fatigue	Rapid; lactate accumulates and an oxygen debt builds up	Slow
Main source of ATP	Glycolysis (anaerobic respiration)	Electron transport chain (aerobic respiration)
Structure of fibres	Thicker in diameter; pale in colour	Thin cross-section; red in colour due to dense mitochondria, lots of blood capillaries and myoglobin
Location	Tend to be concentrated in arms and legs	More widespread; postural muscles attached to core/spine tend to be slow twitch

Exam practice

1 The following diagram shows a neuromuscular junction and a myofibril.

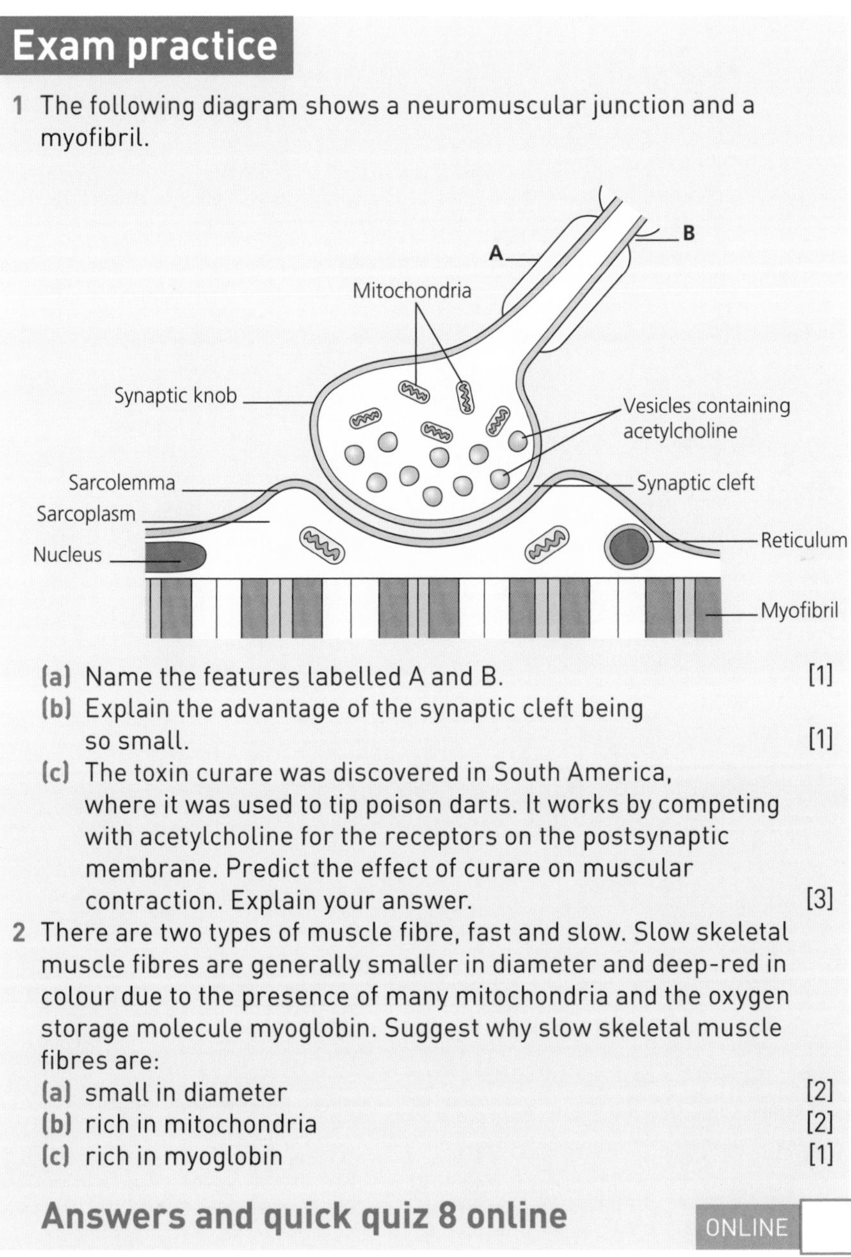

(a) Name the features labelled A and B. [1]

(b) Explain the advantage of the synaptic cleft being so small. [1]

(c) The toxin curare was discovered in South America, where it was used to tip poison darts. It works by competing with acetylcholine for the receptors on the postsynaptic membrane. Predict the effect of curare on muscular contraction. Explain your answer. [3]

2 There are two types of muscle fibre, fast and slow. Slow skeletal muscle fibres are generally smaller in diameter and deep-red in colour due to the presence of many mitochondria and the oxygen storage molecule myoglobin. Suggest why slow skeletal muscle fibres are:

(a) small in diameter [2]

(b) rich in mitochondria [2]

(c) rich in myoglobin [1]

Answers and quick quiz 8 online

ONLINE

Summary

By the end of this chapter you should be able to understand:

- The basic structure of a motor neurone.
- How a resting potential is established as a result of unequal ion distribution.
- The ionic events that lead to the action potential.
- The all-or-nothing principle.
- The nature and importance of the refractory period.
- Factors affecting the speed of conductance.
- The structure of a synapse and neuromuscular junction.
- The events involved in transmission across a synapse and neuromuscular junction.
- How to interpret the action of a given drug.
- The gross and microscopic structure of skeletal muscle, especially myofibrils.
- The roles of actin, myosin, tropomyosin, calcium ions and ATP in muscle contraction.
- The role of ATP and phosphocreatine in providing energy during contraction.
- The structure, location and general properties of slow and fast skeletal muscle fibres.

9 Homeostasis

Principles of homeostasis and negative feedback

The word **homeostasis** means 'steady state', but the conditions within the body of a mammal are controlled within certain limits rather than being kept constant. Overall, there are many homeostatic mechanisms that work together to perform a remarkably effective balancing act, such as:

- maintaining a constant **core temperature** of around 37°C
- maintaining constant **blood pH** at between 7.3 and 7.45

Both conditions need to be kept within narrow limits because of the sensitivity of enzymes. By contrast, **blood glucose concentrations** can vary widely. If levels fall too low, the body's cells become starved of vital fuel. If they rise too high, the water potential of the blood falls too low. However, as long as the levels are kept at between 4 and 9 $mmol\,l^{-1}$, there are usually no problems.

Typical mistake

Students tend to write that 'enzymes are denatured' as the explanation for all enzyme-related problems. However, if the temperature rises too high or the pH is not quite at the optimum level, different enzymes will be affected to a different extent and the result is a metabolic imbalance. It takes more extreme conditions to denature most enzymes.

Negative feedback

REVISED

Negative feedback is the mechanism that keeps things stable (Figure 9.1). The key elements are as follows:

1. A physiological level changes — for example, temperature increases or blood glucose decreases.
2. The change is detected by receptors.
3. A mechanism is activated, via nerve impulses and/or hormones, that reverses the change.
4. The extent of the change is monitored so that when conditions return to normal, the corrective mechanism is switched off.

Homeostasis is the maintenance of constant conditions within the body.

Negative feedback is a mechanism for stability, often referred to as 'detection-correction'. Deviations from a set point are returned to the norm.

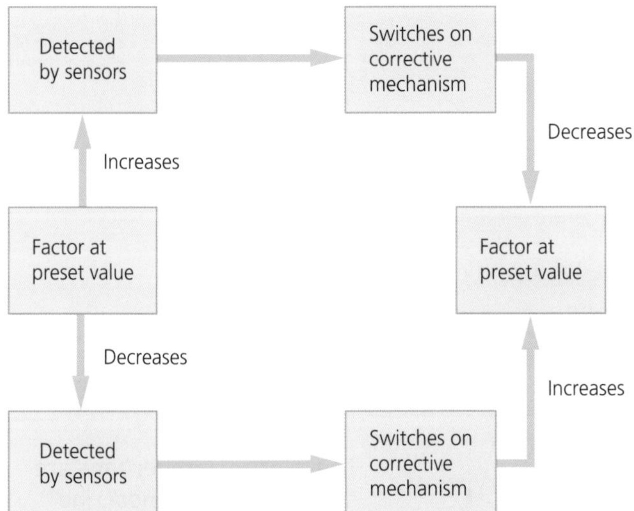

Figure 9.1 Negative feedback is often described as a 'detection-correction' mechanism because deviations from a set point are returned to the norm

Positive feedback

REVISED

Positive feedbacks are mechanisms for change. It is often said that positive feedback is an emergency in which normal homeostatic mechanisms have broken down. For example, when core body temperature exceeds a critical level, enzymes work faster, which generates more heat, which makes the enzymes work even faster. However, there are many situations in which positive feedback brings about a necessary change, such as blood clotting, ovulation and childbirth.

Positive feedback involves a departure from the set point, bringing about changes that produce further change. Think of it as a 'vicious circle' but don't call it that in exams.

Now test yourself

TESTED

1 What is the key difference between positive and negative feedback?

Answer on p. 206

Control of blood glucose concentration

Glucose is absorbed into the bloodstream from the gut following carbohydrate digestion. After a meal there is usually more glucose than the body immediately requires, so the excess is stored as **glycogen** (Figure 9.2), with large amounts being stored in the **liver** and **muscles**.

Glycogen is a storage carbohydrate found in animals. It is a branched polymer of glucose that can be built up and broken down quickly according to the demands of the body.

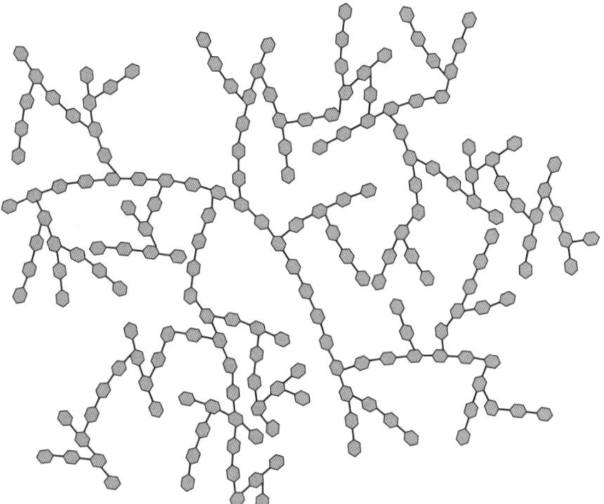

Figure 9.2 A glycogen molecule

Three key processes in glucose control are:

- **glycogenesis** — the synthesis of glycogen
- **glycogenolysis** — the breakdown of glycogen
- **gluconeogenesis** — when glycogen stores run low, new glucose can be generated by the conversion of amino acids and lactate

The role of insulin and glucagon

REVISED

The **pancreas** (Figure 9.3) has a major role in the control of blood glucose. Small groups of cells, **islets of Langerhans**, contain two types of hormone-producing cell:

- α (alpha) cells produce the hormone **glucagon**
- β (beta) cells produce the hormone **insulin** (Figure 9.3)

The **islets of Langerhans** are small patches of endocrine tissue in the pancreas.

These cells are unusual because they are both receptors and effectors:

- If blood glucose levels rise, this is detected by the β cells, which respond by secreting insulin into the blood. Insulin works by increasing the permeability of cell membranes to glucose (Figure 9.4). In this way, glucose can leave the blood and enter cells, thus lowering levels in the blood.
- If blood glucose levels fall, this is detected by the α cells, which respond by secreting glucagon into the blood. This works by activating the enzymes that break down glycogen, generating glucose, which can pass into the blood and raise levels.

Exam tip

There are six G words in this topic, which can be confusing. Three are substances: glucose, glycogen and glucagon. Three are processes: glycogenesis, glycogenolysis and gluconeogenesis. To remember which process is which, remember *neo* = new, *lysis* = splitting, *genesis* = creation.

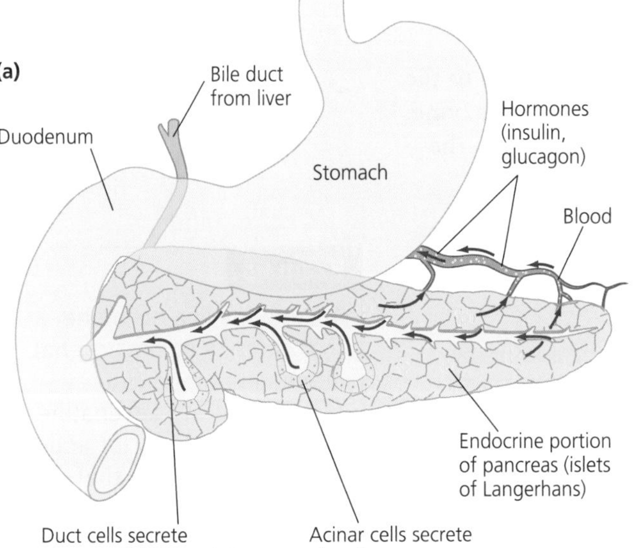

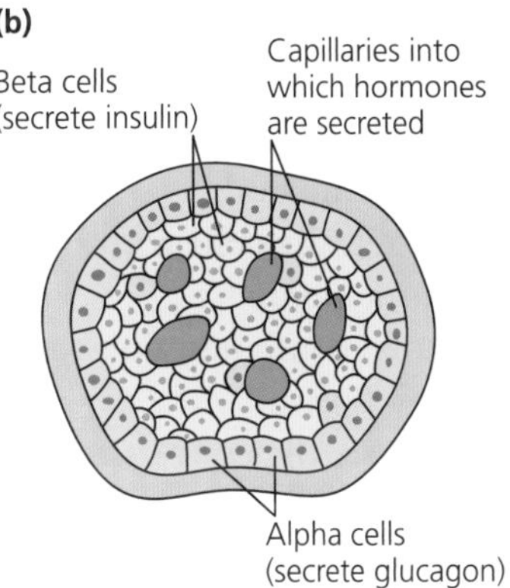

Figure 9.3 (a) Most of the pancreas makes digestive juice, but the islets of Langerhans have a vital role in the control of blood glucose (b) A single islet of Langerhans

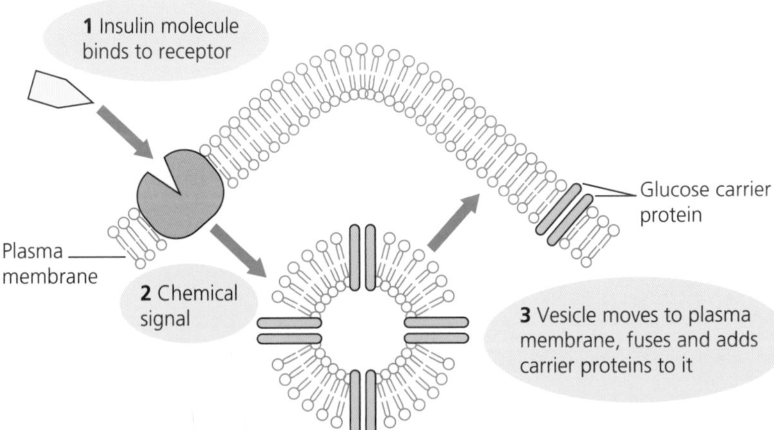

Figure 9.4 Insulin works by causing the cell to add extra glucose transport proteins to the cell surface membrane. Without insulin, most of the glucose transport proteins remain dormant in vesicles in the cytoplasm

How hormones work

REVISED

There are two ways a hormone can work:

- Most hormones, including adrenaline (Figure 9.5), glucagon and ADH, work by the **second messenger model**. They do not enter the cells, but by binding to specific hormone receptors they alter the activity inside the cells.
- **Steroid hormones** are more direct. They are lipid soluble and can pass straight through membranes into the cell and change what happens in the nucleus, usually by activating specific genes. There is more about steroids on page XX–XX.

The second messenger model

A second messenger is a substance found inside a cell that responds to the presence of a hormone outside the cell by activating a particular enzyme:

1. The hormone is released by an endocrine gland and circulates in the blood to all parts of the body.
2. The hormone targets the cells with the correct receptor proteins in their cell surface membrane.
3. When the hormone combines with the receptor, an enzyme is activated in the cytoplasm.
4. The enzyme catalyses the conversion of ATP into cyclic adenosine monophosphate (cAMP). The first messenger was the original hormone. As it cannot get into the cell, a second messenger — cAMP — is needed to cause the desired effect inside the cell.
5. The cAMP activates certain enzymes. In the case of glucagon and adrenaline, it is the enzymes that catalyse the breakdown of glycogen.

This is an example of a **cascade reaction**, where a small amount of hormone can be amplified to bring about a significant effect. It is a chain of chemical events and each chemical can activate many more.

Exam tip

Adrenaline is a hormone that is made by the adrenal glands in response to stress. Overall, adrenaline prepares the body for action in a variety of ways that include raising heartbeat, blood pressure and blood glucose levels. It also dilates the bronchioles in the lungs.

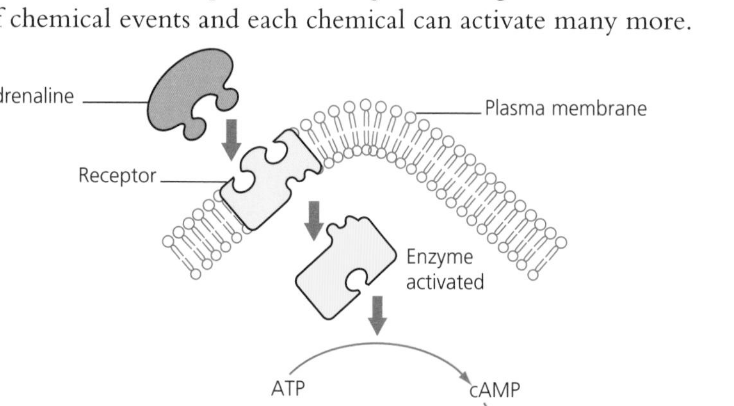

Figure 9.5 The second messenger model of adrenaline action. Glucagon works in much the same way, although the shape of the hormone and the receptor are different

Diabetes

REVISED

There are two types of **diabetes mellitus**: type 1 and type 2.

In **Type 1**, sufferers cannot make insulin and are usually treated by injections of the missing hormone. This is a lifelong condition that often starts in early childhood. The lack of insulin causes glucose to accumulate in the blood because it cannot pass into the cells quickly enough. The cells become starved of fuel and have to respire lipid as an alternative. Classic symptoms are:

- excessive thirst — glucose lowers the water potential of the blood
- excessive urination due to higher fluid intake
- glucose in the urine — the kidneys are unable to re-absorb all the glucose
- weight loss because the cells respire lipid reserves
- breath smelling of ketones (a fruity smell) — a by-product of lipid metabolism

Type 1 diabetics have to monitor their glucose levels regularly. They get used to maintaining sensible blood glucose levels by injecting a mixture of slow- and fast-acting insulin and matching it to their sugar/carbohydrate intake.

In **Type 2** diabetes, sufferers make insulin but it may not be enough or the cells stop responding to it — a problem called **insulin resistance**. Type 2 is also called late-onset diabetes and is associated with being overweight. It is usually treated by diet and exercise to control weight.

There are about 2 million diabetics in the UK, of which about 85–90% have type 2. It is difficult to be precise because many individuals with type 2 are borderline or undiagnosed.

Control of blood water potential

A vital aspect of homeostasis is the control of the water potential of the blood and body fluids. This is called **osmoregulation** and the key organs are the **kidneys**. The kidneys have three functions:

- control of the water potential of the blood
- control of the volume and pressure of the blood
- removal of metabolic waste, such as urea

In this specification we are only concerned with the first function.

Blood is delivered to the kidneys in the **renal arteries** and leaves in the **renal veins**. The **ureters** take the urine from the kidney to the bladder (Figure 9.5).

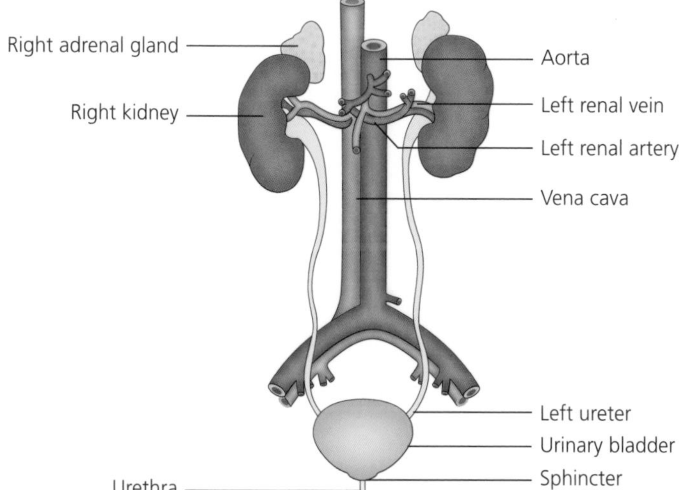

Figure 9.6 The human urinary system. One tube enters each kidney: the renal artery. Two tubes come out: the renal vein carrying filtered blood, and the ureter carrying urine

Overview of kidney function

REVISED

Each kidney is composed of about a million tightly packed tubules (little tubes) called **nephrons**, surrounded by a network of blood vessels (Figures 9.7 and 9.8). Blood entering a nephron is filtered under pressure — a process called **ultrafiltration**. This is similar to the formation of tissue fluid (year 1), and the fluid produced — called **filtrate** — has the same composition as tissue fluid. If the kidneys simply excreted filtrate the body would become dehydrated very quickly, and lose a lot of essential substances. Therefore, once the filtrate has been made, the kidney must reabsorb the substances it needs, and excrete the substances it does not need.

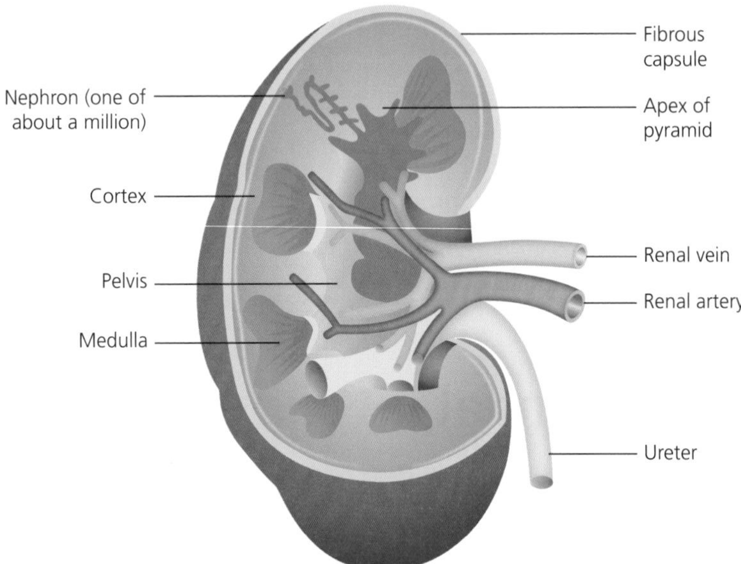

Figure 9.7 A kidney is basically a collection of nephrons and blood vessels. This section through a kidney shows the position of one nephron

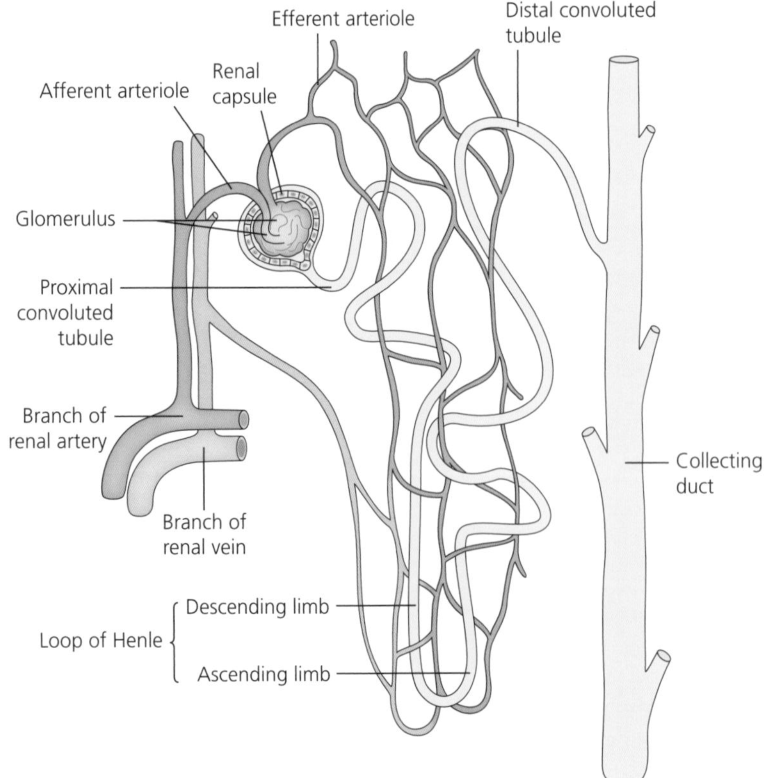

Figure 9.8 A single nephron and its blood supply

The formation of glomerular filtrate

Blood enters the **renal capsule** (Figure 9.9) — sometimes called the **Bowman's capsule** — where the arteriole branches out into a knot of capillaries called the **glomerulus**. The afferent vessel going in is wider than the efferent vessel coming out, which increases the hydrostatic pressure in the glomerulus. As a result, blood is forced against a filter that consists of three layers (Figure 9.10):

- the blood vessel walls
- the basement membrane (a thin mesh of protein that supports the capillary wall). This is the finest part of the filter. Only molecules smaller than about 68 000 daltons can pass through.
- the cells lining the Bowman's capsule, known as **podocytes** ('cells with feet')

Ultrafiltration is a non-selective process that removes a proportion of all of the components of blood plasma that are small enough to pass through. This includes water, ions, glucose, amino acids and urea.

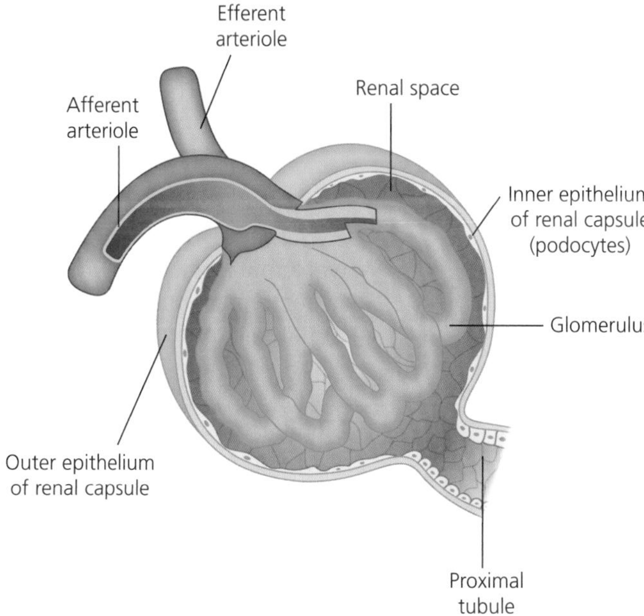

Figure 9.9 The renal capsule

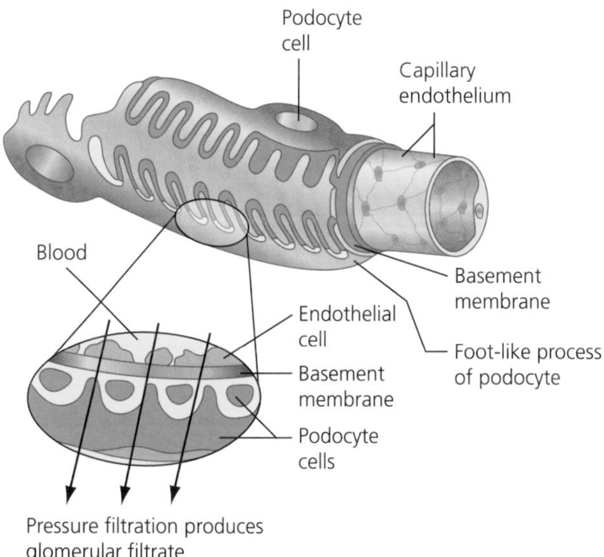

Figure 9.10 The three components of the filter in the renal capsule

The proximal convoluted tubule

This is the main region of reabsorption. Active transport mechanisms in the cells of the proximal tubule reabsorb a lot of the glucose, amino acids and some ions. This is **obligate** reabsorption, which means that it must happen, whatever the conditions. The reabsorption of substances into the blood lowers the water potential, so water leaves the tubule by osmosis. In this way most of the filtrate is returned to the blood.

Proximal means 'near' and **convoluted** means 'twisted'.

Now test yourself

TESTED

2 From the following list:

sodium ions; amino acids; chloride ions; glucose; water; urea; proteins; potassium ions

Choose the substances (one or more) that:

(a) are usually too big to be filtered
(b) will always be reabsorbed (so there is none in the urine)
(c) will always be excreted
(d) might be reabsorbed or excreted depending on the needs of the body at the time

Answer on p. 206

The loop of Henle

This part of the nephron is a hairpin-shaped loop whose function is to create a region of high salt concentration — and therefore low water potential — through which the collecting duct must pass. The loop has evolved to allow organisms to excrete while still conserving water, which they do by making **hypertonic** urine. Put simply, when water is scarce, mammalian kidneys can make really concentrated urine. That is, urine with a lower water potential than body fluids.

Hypertonic means 'lower water potential'. Seawater is hypertonic to human blood. Hypotonic means 'greater water potential'. Distilled water is hypotonic to human blood. Isotonic means 'same water potential'. A 0.9% solution of sodium chloride is isotonic to human blood.

Now test yourself

TESTED

3 Explain why a mammal can never conserve water by stopping urine production completely.

Answer on p. 206

To understand how the loop works, it helps to consider the ascending limb first:

1 As filtrate flows up the ascending limb, an active transport mechanism pumps sodium ions out, resulting in a high concentration of ions and a low water potential in the surrounding fluid. The ascending limb is impermeable to water, otherwise water would simply follow the solute by osmosis and the loop would not be able to function.
2 As fluid passes down the descending limb, the low water potential of the surrounding fluid causes water to leaves by osmosis. Water passes through specific membrane proteins called **aquaporins**. The longer the loop, the more water flows out of the filtrate and the lower the water potential of the filtrate at the apex.

This system is called a **countercurrent multiplier** because the fluid in the two limbs flows in opposite directions, and the overall effect is multiplied by the length of the loop. The longer the loop, the more steps 1 and 2 above have a chance to build up a low water potential around the apex of the loop, deep in the medulla of the kidney.

The overall point of the loop is that it creates a region of high salt concentration, and therefore a very low water potential, through which the collecting duct must pass. Animals that need to conserve water have especially long loops.

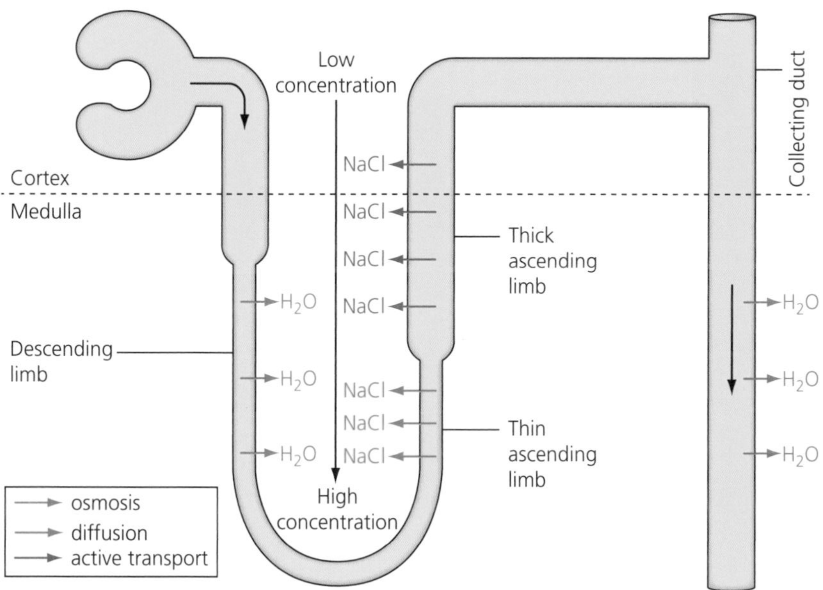

Figure 9.11 The movement of water and salt in the loop of Henle

The distal convoluted tubule (DCT) and collecting ducts

The distal convoluted (far, twisted) tubule and the collecting duct are the regions of the nephron that can re-absorb water if necessary. The function of the proximal tubule is pretty consistent and happens whatever the conditions, but the activities of the DCT and the collecting duct depend on whether or not there is a need to conserve water.

The control of water potential

The control of water potential in the body is known as **osmoregulation**. When water loss exceeds water intake, for example when we sweat and don't drink much, we begin to be dehydrated. The fall in the water potential of the blood is detected by **osmoreceptor** cells in the **hypothalamus** (Figure 9.11), which responds by making **ADH** — **anti-diuretic hormone**. ADH is then transported to the pituitary, where it is secreted from the posterior (rear) lobe.

Without ADH, the cells that make up the DCT and collecting duct are impermeable to water. As a result, all of the water in the filtrate passes through into the urine. In order to conserve water, the permeability of these regions to water must be increased, and this is done by membrane proteins called **aquaporins**, which are stored in vesicles under the surface of the cells.

ADH works — via the second messenger model — by causing vesicles to come to the surface of the cell, so the aquaporins join the membranes of the DCT and collecting duct. This allows water to leave the filtrate by osmosis, due to the low water potential created by the salt concentration. The blood vessels — the **vasa recta** — that pass through this region absorb the water.

Exam tip

The term 'diuresis' means urine production, so ADH is the 'stop-you-peeing hormone'. Don't call it that in the exam.

Now test yourself

TESTED

4 Explain how the action of ADH is similar to the action of insulin (page 152).

Answer on p. 206

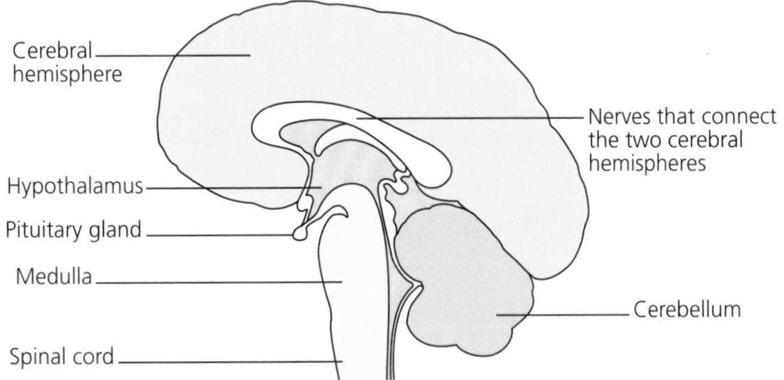

Figure 9.12 A section through the centre of the brain, showing the position of the hypothalamus and pituitary gland. Generally, the hypothalamus controls the activities of the pituitary gland

Now test yourself

TESTED

5 Explain how the control of water potential is an example of negative feedback.

Answer on p. 207

Exam practice

1 If an individual is suspected of being diabetic, a glucose tolerance test can be done. The patient has to fast for 12 hours (usually overnight) before being given a drink containing 75 g of glucose. The blood glucose levels are then monitored for 3 hours. The following graph shows the results for three individuals, A, B and C.

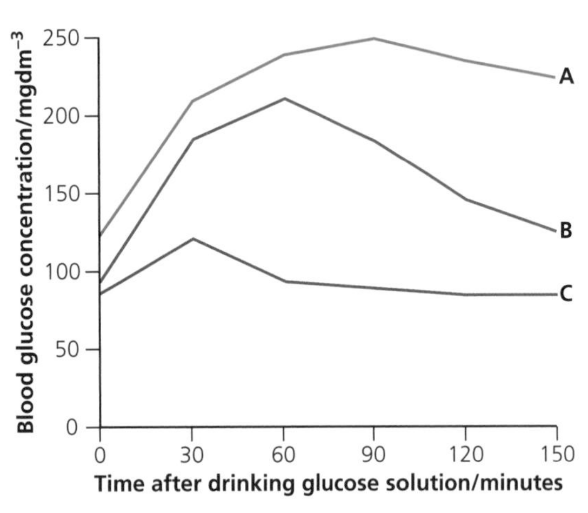

(a) Suggest why patients have to fast for 12 hours before the test. [1]
(b) Explain why there is a rise in all three graphs for the first 30 minutes. [1]
(c) Explain how a healthy individual responds to a rise in blood glucose. [3]
(d) Individual C is a type 1 diabetic. Give *three* differences between the curves for individuals A and C. [3]
(e) Individual B is a borderline type 2 diabetic. Suggest what treatment was suggested. [1]

2 (a) In the kidney nephron, filtrate is formed by ultrafiltration, which requires hydrostatic pressure.
(i) Specifically, which part of the body creates hydrostatic pressure? [1]
(ii) Explain how the blood vessels in the renal tubule further increase this pressure. [1]

In the following question parts, each answer builds on the previous one. In the real exam it is unlikely that you will be set a question like this because one mistake could lose you a disproportionate amount of marks.

On average, the heart pumps about 5 litres of blood per minute. This is known as the cardiac output.

(b) 21% of the cardiac output flows to the kidneys. Calculate the blood flow to the kidneys in cm^3 per minute. [2]

(c) A large proportion of blood is taken up by cells, mainly red cells. This value, known as the haematocrit, is typically about 40%. The rest is plasma. Calculate the volume of plasma reaching the kidneys in cm^3 per minute. [2]

(d) The glomerular filtration rate (GFR) is the amount of filtrate formed per minute. Typically, it is 19% of the plasma that flows through the renal capsules. Calculate the GFR in $cm^3 min^{-1}$. [2]

(e) On average, the human kidney makes about 1 cm^3 of urine per minute. Calculate the percentage of the filtrate that is reabsorbed into the blood. [2]

Answers and quick quiz 9 online

ONLINE

Summary

By the end of this chapter you should be able to understand the following:

- Homeostasis involves control systems that maintain the internal environment within certain limits.
- Negative feedback is a 'detection-correction' mechanism that keeps the body's internal conditions within certain limits.
- Positive feedback is a mechanism for change.
- The factors that affect blood glucose levels include diet, exercise and stress.
- The role of insulin and glucagon in controlling blood glucose levels.
- The effect of adrenaline on glycogen breakdown and synthesis.
- The second messenger model of hormone action.
- The causes, symptoms and treatment of types 1 and 2 diabetes.
- The kidneys are the organs of osmoregulation — the control of the water potential of the blood and tissue fluids.
- The kidneys work on two basic processes: ultrafiltration and secretion/reabsorption.
- Ultrafiltration is filtration under pressure and takes place in the renal capsule.
- Most of the filtrate is reabsorbed in the PCT.
- Water is excreted or conserved according to the needs of the body.
- A negative feedback system involving the hypothalamus and ADH controls the water potential of the blood and body fluids.

10 Inheritance

The **genotype** is the genetic constitution of an organism: the alleles it has inherited. It is written using notation such as **Aa** or **AA**. The **phenotype** refers to the observable characteristics of an organism, and results from a combination of the **alleles** it has inherited and its environment.

Alleles are alternative versions of the same gene (Figure 10.1). For example, a particular species of plant might have a gene for flower colour that has two alleles: one codes for purple flowers and one for red flowers. New alleles are created by the process of **mutation**.

A specific gene will always occur at the same position on a chromosome. This is called its **locus**.

Organisms usually have two copies of each gene because chromosomes come in pairs. If the alleles at a specific locus are the same, they are **homozygous**, written as **AA** or **aa**, for example. If they are different, they are **heterozygous** (**Aa**).

The **genotype** is the alleles an organism possesses. Genotypes are usually written using letters to show the different alleles, such as Bb or BB.

The **phenotype** is the observable characteristics an organism possesses. For example, blonde hair and left-handed.

An **allele** is an alternative form of a gene. For example, a gene for flower colour could have an allele for red flowers and an allele for white flowers.

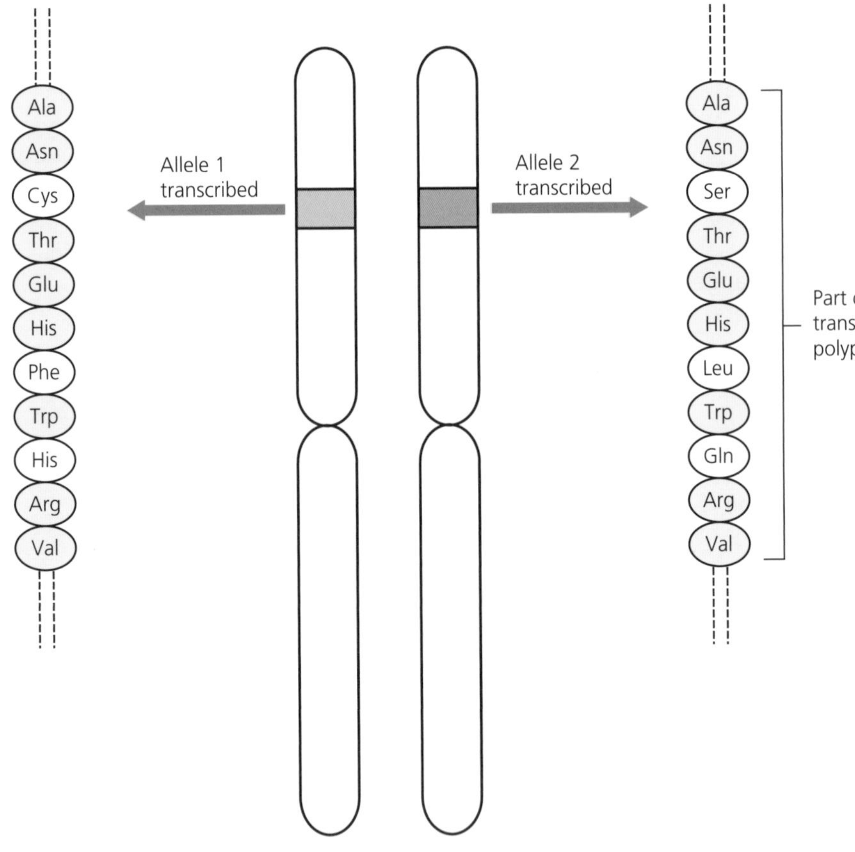

Figure 10.1 Different alleles will have different base sequences and so will code for different amino acids. When assembled, the amino acid chain will fold and bend into a different polypeptide or protein

Dominant alleles are those that, if present, are expressed in the phenotype. **Recessive alleles** are only expressed when no dominant allele is present.

Now test yourself

TESTED

1 Define the term *allele*.

Answer on p. 207

Monohybrid crosses

Monohybrid crosses involve **single genes**, usually with two alleles. In this example we will look at cystic fibrosis, an inherited genetic disorder caused by a single faulty recessive allele. This means that people with the condition must have inherited a faulty allele from both parents (Figure 10.2). The normal **F** allele is dominant, so as long as an individual has one working allele it does not matter what the other one is. The **f** allele simply produces a protein that fails to function — it doesn't do any harm.

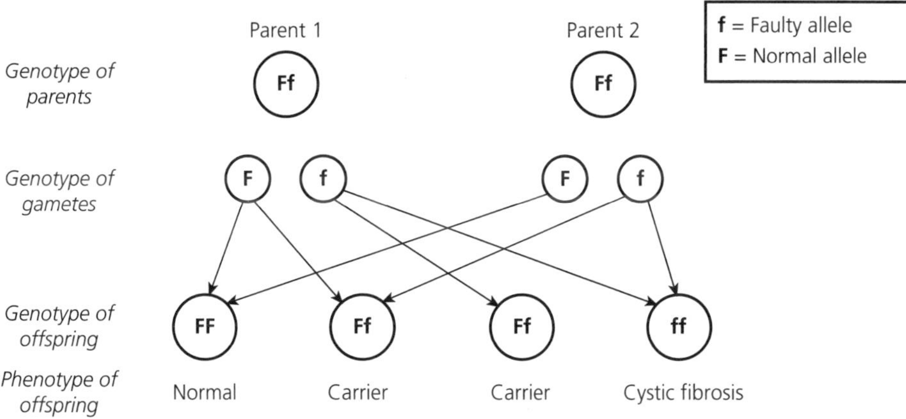

Figure 10.2 Genetic diagram to show the inheritance of cystic fibrosis

Notice that the offspring with **Ff** are labelled 'carriers'. Carriers have one copy of the faulty allele, but do not have the disorder themselves. In this example, both parents are carriers. If both parents are carriers, there is a 1 in 4 or 25% chance that the child will be born with cystic fibrosis. Probability should be expressed as a decimal, in this case 0.25.

Exam tip

Students often lose marks by expressing the results in the wrong form. A 3:1 ratio, 1 in 4, 0.25 and 25% are all the same thing, but make sure you give the answer in the form required by the question.

Now test yourself

TESTED

2 If the parents are **FF** and **Ff**, what is the probability that their first child will be born with cystic fibrosis?

Answer on p. 207

Codominance

REVISED

Codominance is seen where there is no dominant and no recessive characteristic so, if present, both alleles will be expressed in the phenotype. For example, snapdragons have codominant alleles for red flowers (genotype **RR**) and white flowers (genotype **WW**). If both alleles are present, the flowers are pink (genotype **RW**). Both alleles R and W are equally dominant (Figure 10.3).

Codominance is a situation in which both alleles contribute to the phenotype.

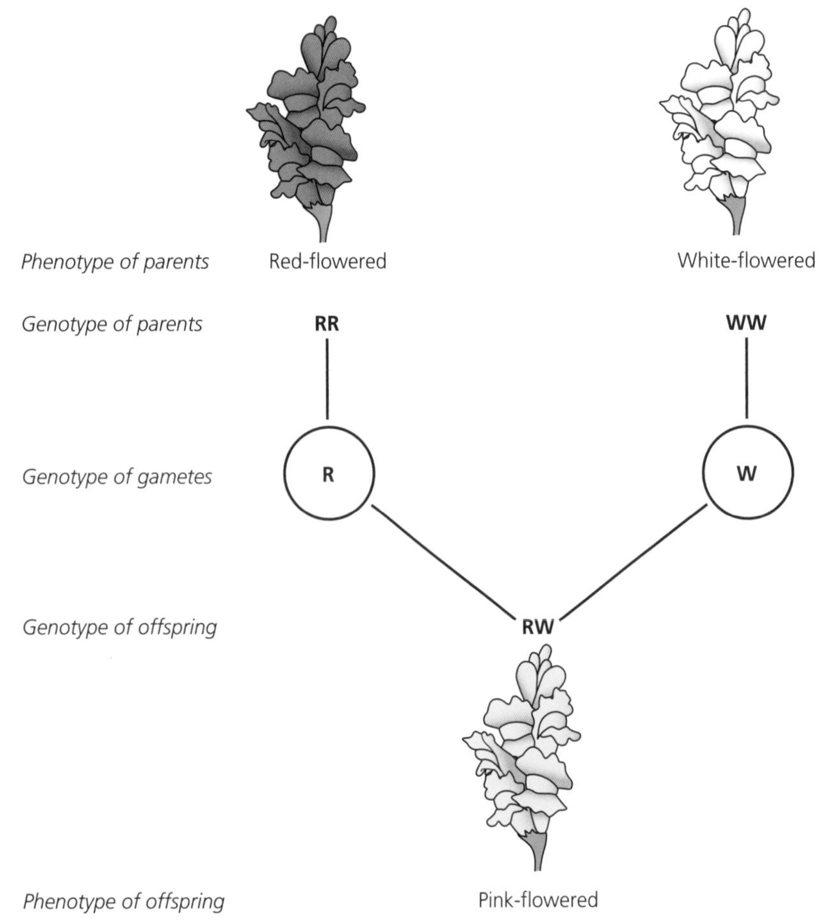

Figure 10.3 Codominance in snapdragons

Now test yourself

TESTED

3 Define the term *codominance*.
4 If two pink snapdragons were bred together, predict the colour and ratios of the offspring.

Answers on p. 207

Multiple alleles

REVISED

Having studied genetics, it is tempting to think that all genes have two alleles, but they do not:

- Some genes have no alleles — there is just one form, often because the gene codes for a vital protein such as an enzyme, and any new alleles are seriously disadvantageous.
- Some genes have two alleles, as we have seen.
- Some genes have multiple (more than two) alleles.

For example, the gene that codes for the human ABO blood group has three alleles, $\mathbf{I^A}$, $\mathbf{I^B}$ and $\mathbf{I^o}$. $\mathbf{I^A}$ and $\mathbf{I^B}$ are codominant over $\mathbf{I^o}$. Each individual has two copies of these alleles, so:

- genotypes $\mathbf{I^AI^o}$ or $\mathbf{I^AI^A}$ will produce blood group A
- genotypes $\mathbf{I^BI^o}$ or $\mathbf{I^BI^B}$ will produce blood group B
- genotype $\mathbf{I^oI^o}$ will produce blood group O
- because of the codominance, genotype $\mathbf{I^AI^B}$ will produce blood group AB

Now test yourself

TESTED

5 A child has a mother with blood group AB and a father with group O. List the possible blood groups of the child.

Answer on p. 207

Family tree diagrams

REVISED

Family tree or pedigree diagrams are popular in exam questions. Figure 10.4 shows the inheritance of albinism in a particular family. Albinos have genotype **aa** — they cannot make normal skin pigmentation.

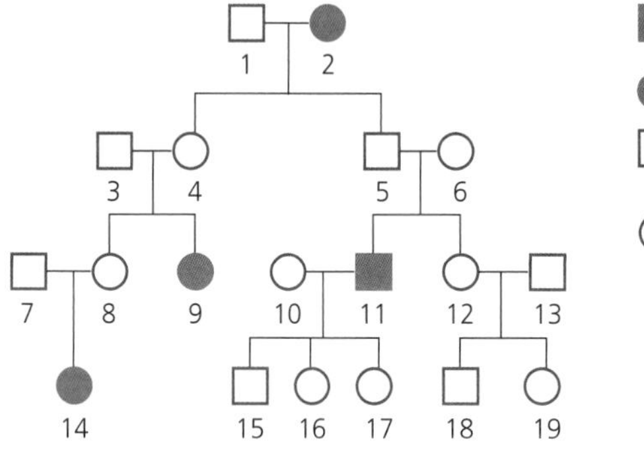

Figure 10.4 Albinism

If **A** is the allele for normal pigmentation and **a** is the albino allele, we can work out the genotypes of many family members. For example, we know that individual 2 must be **aa**, whereas individual 1 could be **AA** or **Aa**.

Now test yourself

TESTED

6 What evidence from Figure 10.4 shows that the albinism allele is recessive?

Answer on p. 207

Sex-linked characteristics

REVISED

Figure 10.5 shows a complete set of human chromosomes. There are 23 pairs in total, 22 pairs of **autosomes** and one pair of **sex chromosomes**. The autosomes are known as **homologous pairs** because they have the same genes at the same loci, although they may or may not have the same alleles. Females have two large X chromosomes, which are homologous. Males have one X and one smaller Y chromosome, which are not homologous.

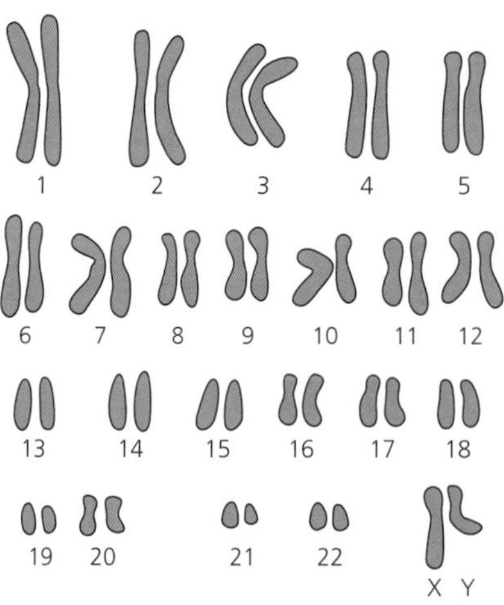

Figure 10.5 A karyotype — the full set of human chromosomes

Exam tip

Remember from cell division in year 1 that homologous chromosomes pair up with each other during the first division of meiosis. The process of **crossover** swaps blocks of genes between the homologous chromosomes, and so makes new allele combinations. This is a central idea in dihybrid inheritance (page XX).

If a gene is found on the sex chromosomes, it is said to be **sex linked**. Currently, it is estimated that the X chromosome carries just under 2000 genes. These are said to be X linked. There are far fewer Y-linked genes — probably less than 100.

The key to understanding sex-linked inheritance is that males have only one copy of X-linked genes, so that allele is always expressed — there is no second copy to mask its effects. Males cannot be carriers. If they are unlucky enough to have a faulty allele, they will have the disease or condition that the allele causes. Examples of sex-linked disorders include haemophilia, colour blindness and Duchenne muscular dystrophy. All are X-linked, so males are more likely to have them.

Now test yourself TESTED ☐

7 What is a sex-linked gene?
8 Explain why males are more likely than females to suffer from sex-linked disorders.

Answers on p. 207

Haemophilia

With sex-linkage, we show the chromosome and the allele. Haemophilia is an inherited, sex-linked genetic disorder that impairs the body's ability to control blood clotting and coagulation. **H** = normal allele; **h** = faulty allele. So, females have three possible genotypes:

- X^HX^H = normal, healthy
- X^HX^h = healthy but a carrier
- X^hX^h = haemophiliac

Males have two possible genotypes — there is no allele on the Y chromosome:

- X^HY = normal, healthy
- X^hY = haemophiliac

In order for two healthy parents to have a haemophiliac son, the mother must be a carrier, so we know what the parents' genotypes must be.

Parental genotypes:	X^HX^h	X^HY		
Gametes:	X^H X^h	X^H Y		
Offspring genotypes:	X^HX^H	X^HX^h	X^HY	X^hY
Offspring phenotypes:	healthy female	carrier female	healthy male	haemophiliac male

Exam tip

Students often add random X and Y chromosomes when there is no need. If the question is about sex linkage, it will clearly say so. If not, it isn't.

Exam tip

Not all species have XX females and XY males. Chickens, for example, are the other way around. The basic genetics is similar, but the sexes are reversed. In this case, it is the females that are much more likely to suffer from sex-linked characteristics.

Now test yourself TESTED ☐

9 Is it possible to have a female haemophiliac? Explain your answer.

Answer on p. 207

Sex-linked pedigree diagrams

When trying to make sense of pedigree diagrams involving sex-linked traits, the golden rule is that all males get their Y chromosome from their father — otherwise they would not be males — and so their X chromosome must have come from their mother. Figure 10.6 shows the inheritance of sex-linked colour blindness in one family. Allele **B** for normal vision is dominant to the allele **b** for colour blindness.

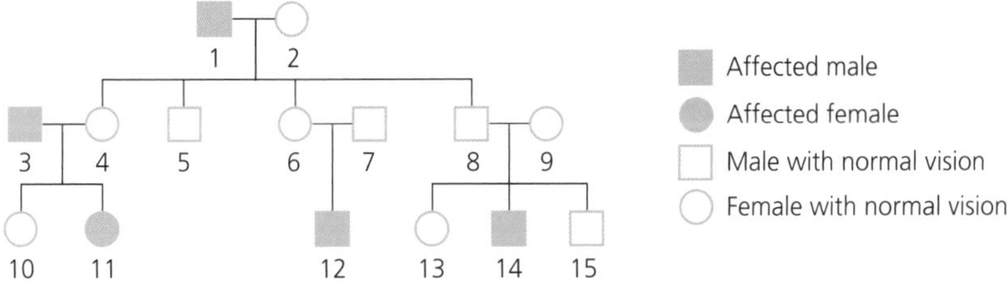

Figure 10.6 Inheritance of red/green colour blindness in one family

Now test yourself

TESTED

10 Work out the genotypes of each individual in Figure 10.6 and give a brief reason for each. Use a table like this for your answer.

Individual	Genotype	Reason
1	X^bY	He is a colour-blind male
2		
3		

Answer on p. 207

Typical mistake

In questions involving family tree (pedigree) diagrams, students often try to state general rules such as 'it is sex linked because more males are affected'. It is much better to be specific, by referring to individuals by number.

Dihybrid inheritance

Dihybrid inheritance concerns the inheritance of two genes. To start with, we look at two genes that are found on different chromosomes. When the genes are found on the same chromosome they are said to be **linked**. We will look at that later.

There are usually three vital parts to dihybrid cross questions:

1 **The parents' genotypes**. These will either be given to you, or you may have to work them out from a description. The genotype tells us some important information. For example, the genotype **AaBb** tells us that:
   - there are two genes
   - with at least two alleles
   - on different chromosomes
   - and this individual is heterozygous for both genes

2 **The gametes**. This is where your knowledge of meiosis can be applied. You need to remember what is meant by **independent segregation**: one chromosome from each homologous pair — and therefore the allele on that chromosome — passes into the gamete with any one from the other pairs.
   It is a good idea to make the gametes clear by putting a circle round them.
   - So, a genotype of **AaBb** can give gametes of **AB**, **aB**, **Ab** or **ab**.
   - A genotype of **AABb** can give gametes of **AB** or **Ab**.
   - And a genotype of **AABB** can give gametes of just **AB**.

Exam tip

F1 stands for 'first filial generation', while F2 refers to the 'second filial generation'. Think of them as the children and the grandchildren.

3 **Work out the outcomes of all possible fertilisations**. Fertilisation is always random, so any male gamete can combine with any female gamete. The genotypes are simply made by adding the gametes together, so that gametes AB and Ab, for example, fertilise to give the genotype of **AABb**.

Exam tip

With dihybrid inheritance there will always be one, two or four different gametes, never three.

An example of non-linked dihybrid inheritance

REVISED

In a particular species of rat the allele **A** for brown hair is dominant to allele **a** for albino hair. A second allele, **B**, the allele for short hair, is dominant to **b**, the allele for long hair. Dihybrid inheritance shows the importance of meiosis in producing many different genotypes by creating new allele combinations.

Let's look at a cross between two homozygous rats: one brown, short-haired male and an albino, long-haired female.

Parent	Male	Female
Genotype	**AABB**	**aabb**
Gametes	(AB)	(ab)
F1 genotypes	All **AaBb**	
Phenotypes	All brown, short haired	

The importance of meiosis is seen when the F1 generation is crossed. Both parents are now heterozygous for both alleles, so meiosis produces four different gametes.

Parent	Male	Female
Genotype	**AaBb**	**AaBb**
Gametes	(AB) (Ab) (aB) (ab)	(AB) (Ab) (aB) (ab)

A Punnet square is a good way to organise the information:

	Female gametes			
Male gametes	**AB**	**Ab**	**aB**	**ab**
AB	**AABB**	**AABb**	**AaBB**	**AaBb**
Ab	**AABb**	**AAbb**	**AaBb**	**Aabb**
aB	**AaBB**	**AaBb**	**aaBB**	**aaBb**
ab	**AaBb**	**Aabb**	**aaBb**	**aabb**

It is often a good idea to keep a tally of phenotypes to get the final ratio:

Phenotype	Tally	Total
Brown, short hair	~~IIII~~ IIII	9
Brown, long hair	III	3
Albino, short hair	III	3
Albino, long hair	I	1

The 9:3:3:1 ratio is only seen when both parents are heterozygous for both alleles.

Now test yourself

TESTED

11 Predict the genotypes, phenotypes and ratios that would result from the following crosses.
(a) Aabb × aaBb (b) AaBb × AABb

Answer on p. 207

Linkage

When the two genes have their loci on the same autosome, independent assortment cannot separate them — they are linked. When alleles occur on the same chromosome, they will be inherited together unless they are separated by crossing over in meiosis. For example, let's look again at an organism with the genotype **AaBb**, but this time the loci for the two genes are on the same chromosome. If one chromosome carries the alleles **A** and **B**, and the other one carries **a** and **b**, that pair of alleles will always be inherited together unless they are separated by crossing over in meiosis. In other words, gametes with the genotype **AB** and **ab** will be much more common than the recombinant gametes carrying **Ab** and **aB**. As a consequence, genotypes **AABB** and **aabb** — like the original parents — will be much more common too (Figure 10.7).

The loci of the linked genes are important here. If the loci are at opposite ends of the chromosome, there is a good chance that **chiasmata** will form between them, so that recombinants (**Ab** and **aB**) will be relatively common. If the loci are close, recombinants will be rare. You can tell a lot about the loci of different genes by the frequency with which they are separated by crossover. Scientists use this information to make chromosome maps.

If the loci are close together, there is much less chance that they will be separated by crossover. Linkage groups are clusters of genes that tend to be inherited together because their loci are very close.

Exam tip

It is important that you practise dihybrid crosses before the exam. The first ones you do will take a while, but you will quickly learn short cuts.

Typical mistake

Questions that require you to work out dihybrid crosses commonly have 3 marks. Students lose marks either because they fail to work out the gametes clearly, they get the F1 genotypes wrong, they interpret the genotypes incorrectly or they get the ratios wrong.

Chiasmata are the points of connection that form between two homologous chromosomes during crossover.

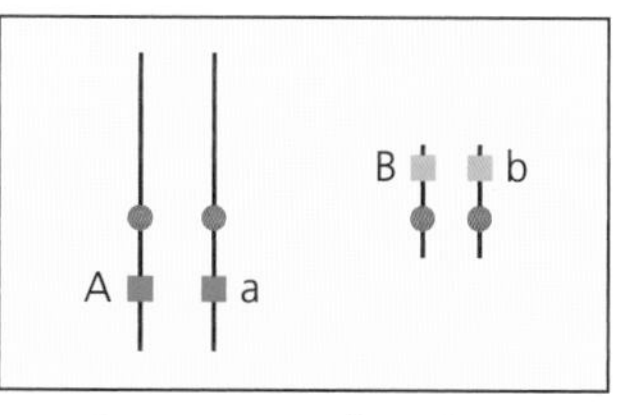
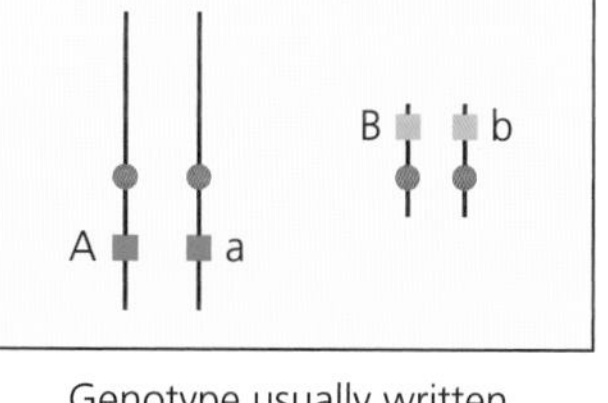
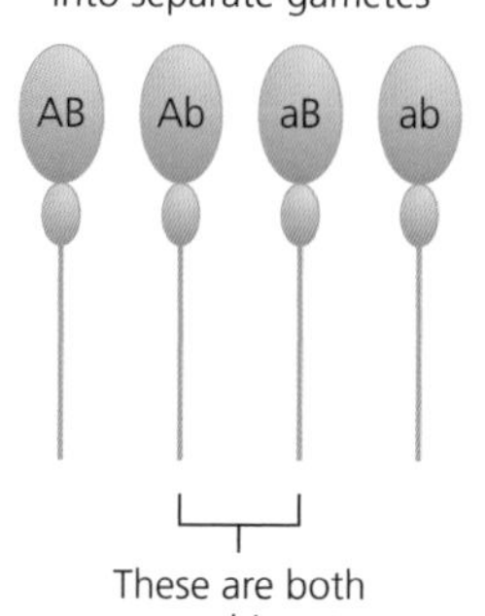
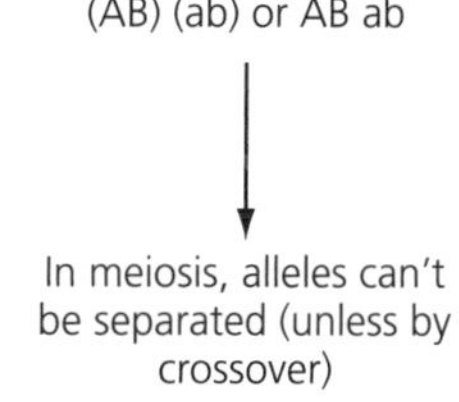
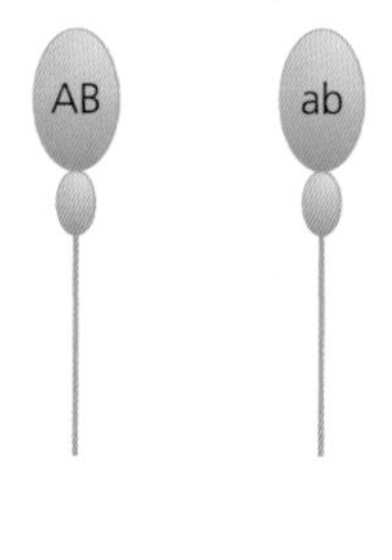
Unlinked alleles
A and B are not linked, i.e. they are on different chromosomes
B
b
A
a
Genotype usually written AaBb
In meiosis, independent assortment separates alleles into separate gametes
AB
Ab
aB
ab
These are both recombinants
Linked alleles
A and B are linked, i.e. present on the same chromosome
A
B
a
b
This is usually written as (AB) (ab) or AB ab
In meiosis, alleles can't be separated (unless by crossover)
AB
ab
There are no recombinants

Figure 10.7 The effect of linkage. (a) The situation when the two genes are not linked — the four different gametes are produced in equal numbers. (b) When the genes are linked, there will be far fewer recombinants. Alleles A and B will always pass into the gamete, as will alleles a and b, unless separated by chiasmata during crossover

Now test yourself

TESTED

12 (a) Define the genetic term *linkage*.
(b) Explain why independent segregation cannot separate linked genes.

In the sweet pea plant, the allele for purple flowers is dominant to the allele for red flowers. Another gene codes for the shape of seed — the allele for elongated is dominant to the allele for round. When a pure-breeding purple, elongated plant was bred with a red, rounded individual, you would expect a 9:3:3:1 ratio in the F2 generation. The table below shows the results from one breeding experiment.

Phenotype	Expected numbers for a 9:3:3:1 ratio	Numbers obtained in the F2 generation
Purple, elongated	405	336
Purple, rounded	135	31
Red, elongated	135	28
Red, rounded	45	325
Total	720	720

(c) Explain how the observed numbers support the idea that the genes for flower colour and seed shape are linked.

Answer on p. 207

Epistasis

Epistasis is a situation where the expression of a particular gene depends on the presence of a particular allele of another gene. Epistatic interactions are very common. As an example, imagine the red colour of the petals in a particular plant species is made in two stages:

compound A (no colour/white) → compound B (yellow pigment) → compound C (red pigment)

This metabolic pathway needs two enzymes. Enzyme 1 converts A into B, and enzyme 2 converts B into C. These enzymes are made by two genes, **A**/**a** and **B**/**b**.

- Allele **A** codes for enzyme 1; allele **a** is mutant allele that produces a non-functional enzyme.
- Similarly, allele **B** codes for enzyme 2; allele **b** is a mutant allele that produces a (different) non-functional enzyme.

You can get the basic idea behind epistasis by considering the different genotypes. Those with genotype **AABB** will have red flowers. In fact, any genotype that has both **A** and **B** will produce red flowers. We can write this as **A–B–**.

But, if allele **A** is not present, so the genotype is **aa**, it doesn't matter if allele **B** is present or not, the genotype will be white because enzyme 2 will have no yellow substrate to work on.

Now test yourself

TESTED

13 Using the example of the plant species in the text, give all of the genotypes that will produce:
(a) yellow flowers
(b) white flowers

Answer on p. 207

Use of the chi-squared (χ^2) test in genetics exams

Chi-squared is a simple statistical test that allows you to compare the 'goodness of fit' of observed phenotypic ratios compared with expected ratios. It basically asks: 'are the genes behaving as we think they are, or is there likely to be a different explanation?' For example, in a particular dihybrid cross with no linkage you might be expecting a phenotypic ratio of 9:3:3:1. In genetics, all ratios are averages, so you will rarely get exactly 9:3:3:1. The vital question is: how far from the expected ratio can you go before you can say 'it's probably not 9:3:3:1' and start looking for another explanation, such as linkage? The chi-squared test will give you the *probability* that your results fit your expected ratio.

Example

This example comes from one of Mendel's famous experiments on peas. He crossed pure-breeding (homozygous) pea plants that produce round, yellow peas with pure-breeding plants that produced wrinkled, green seeds. In the F1 generation all the seeds were round and yellow but when the F1 were bred together ('selfed') he expected a 9:3:3:1 ratio in the F2 generation. Out of 400 peas, this is what he got:

Round, yellow peas	Round, green peas	Wrinkled, yellow peas	Wrinkled, green peas
219	81	69	31

Is this a 9:3:3:1 ratio?

In this situation, the **null hypothesis** would be: there is no difference between the observed and the expected — the 9:3:3:1 ratio.

So now we have two sets of values: the ones we actually got — the **observed** — and the values that fit the 9:3:3:1 ratio perfectly — the **expected**. The chi-squared formula is

$$\chi^2 = \sum \frac{[(O - E)^2}{E}$$

Where O is observed, E is expected and Σ (sigma) means 'the sum of all'.

With 400 crosses, a 9:3:3:1 ratio would give us expected values of 225, 75, 75 and 25, and the chi-squared formula will compare those to the observed values. It's a good idea to build the formula step by step, using a table like the one below. We take the difference between the observed and the expected values and square them to get rid of the negative numbers.

Phenotype	Observed (O)	Expected (E)	$O - E$	$(O - E)^2$	E	$\frac{(O - E)^2}{E}$
Round Yellow Peas	219	225	−6	36	225	0.16
Round Green Peas	81	75	6	36	75	0.48
Wrinkled Yellow Peas	69	75	−6	36	75	0.48
Wrinkled Green Peas	31	25	6	36	25	1.44
Total	**400**	**400**				**Σ = 2.56**

So our χ^2 value is 2.56. What does that mean? We also need another value: the number of **degrees of freedom**. This takes into account the number of different sets of data, which in this case is 4 because we have four different phenotypes. The more sets of data there are, the greater the variation is likely to be. The value for degrees of freedom is always one less than the number of sets of data, which in this example is 3 (4 − 1).

The final step is to look up our two values (2.56 with 3 degrees of freedom) in a χ^2 probability table (below).

Degrees of freedom	Probability					
	0.50	0.25	0.10	0.05	0.02	0.01
1	0.45	1.32	2.71	3.84	5.41	6.64
2	1.39	2.77	4.61	5.99	7.82	9.21
3	**2.37**	**4.11**	6.25	7.82	9.84	11.34
4	3.36	5.39	7.78	9.49	11.67	13.28

If we look along the row for 3 degrees of freedom, we can see that our value (2.56) is between the values highlighted in the table. If you look at the column headings, they show a probability between 0.5 (or 50%) and 0.25 (25%). This means that you would expect observed values to differ from the expected values by as much as this between 25% and 50% of the time, and that's very high. For most tests, the critical value for a significant difference is 0.05 (or 5%).

So we have to accept the null hypothesis in this case – there is no significant difference between the observed results and those we would expect if the ratio was 9:3:3:1. Looking at the table above, the threshold χ^2 value for a significant difference would be 7.82, which corresponds to a probability of 0.05,or 5%. This would mean that the probability that the observed and expected results are significantly different is greater than 95%, so we can look for another genetic mechanism, such as epistasis or linkage.

Population genetics

So far we have looked at the genetics of individuals. In this section we look at the genetics of populations. The definition of a population is a group of individuals of the same species that are capable of interbreeding.

The Hardy–Weinberg principle

REVISED

The Hardy–Weinberg principle is a calculation that allows you to:

- calculate **allele frequencies** in a population, starting from simple observations
- make predictions, for example, is a species evolving?

Allele frequency is a measure of how common an allele is in a population. It is expressed as a decimal, so a frequency of 1 means 100%. If a gene has two alleles, **A** and **a**, their frequencies must add up to 1. If the frequency of **A** is 0.7 (70%), we know that the frequency of **a** must be 0.3 (30%). In a population of 100 individuals, there will be 200 alleles, because they are diploid organisms with two copies of each gene. If 140 of the alleles are **A**, 60 must be **a**.

If p is the frequency of the dominant allele, **A**, and q is the frequency of the recessive allele, **a**, then we know that $p + q = 1$, meaning that when you add up all the alleles, you get the whole population.

So imagine a population contains individuals of three genotypes: **AA**, **Aa** and **aa**, where **A** is the dominant allele and **a** is the recessive allele. This gives us the equation:

$$p^2 + 2pq + q^2 = 1$$

where:

- p^2 is the frequency of homozygous dominant individuals (**AA**)
- $2pq$ is the frequency of heterozygous individuals (**Aa**); it's $2pq$ because there are two ways of getting the heterozygous genotype (**Aa** or **aA**)
- q^2 is the frequency of homozygous recessive individuals (**aa**)

When you add them all up, you must get the whole population (1, or 100%)

Hardy–Weinberg calculations

In this example, we will look at coat colour in mice. This is controlled by one gene and two alleles. Allele **A** codes for agouti fur (the normal coloration of wild mice), which is dominant over **a**, which makes no pigment. So, **AA** and **Aa** mice are agouti, whereas **aa** mice are albino.

Starting with the simple observation that 16% of mice are albino, we can work out how many of them are **AA** and how many are **Aa**, despite the fact that the agouti mice all look the same. The individuals with the homozygous recessive genotype can be recognised — they show the recessive phenotype. So, in a population where 16% of individuals show the recessive feature, this is a frequency of 0.16. Therefore:

$$q^2 = 0.16$$

So:

$$q = \sqrt{0.16}$$
$$= 0.4$$

We know that $q = 0.4$, so p must be 0.6 because they add up to 1. Knowing the values of p and q, we can calculate the frequencies of the different genotypes in the population:

The frequency of the homozygous (**AA**) dominant (p^2) is:

$$0.6^2 = 0.36$$

The frequency of the heterozygous (**Aa**) genotype ($2pq$) is:

$$2 \times 0.6 \times 0.4 = 0.48$$

In Hardy–Weinberg questions, many students get confused and end up scribbling formulae and numbers all over the place. It may help to draw a grid similar to the one in Table 10.2, either in rough or at the side of the question. This will help you to focus your thoughts on the information you have been given.

Table 10.2 Sample grid

Genotype	Phenotype	Hardy–Weinberg equation	Value
AA	Agouti	p^2	
Aa	Agouti	$2pq$	
aa	Albino	q^2	0.16

Exam tip

Hardy–Weinberg questions often give you nice numbers to work with. If you find yourself with values like 0.16, 0.36 or 0.48, smile — it's easy to find the square root.

Exam tip

Some questions may give you the allele frequencies. This makes life easier because you don't have to work out the square root of the q^2 value, but the simplicity of the calculations seems to confuse many students.

Now test yourself

TESTED

14 The frequency of one allele is 0.78 and the frequency of a different one is 0.65. How do we know they cannot be alleles of the same gene?

Answer on p. 207

Gene pools, inbreeding and outbreeding

REVISED

A population with a large **gene pool** contains a lot of individuals with different genotypes. There are many alleles and a lot of **outbreeding**, in which genetically different individuals reproduce together. If conditions become unfavourable, there is a good chance that some individuals will have favourable genotypes that will allow them to survive. Faulty alleles are rarely paired up, so few individuals suffer from genetic disease.

A **gene pool** is all the alleles present in a particular population at a given time.

Revision activity

Rewrite this above paragraph, reversing each point to illustrate the other extreme.

Now test yourself

TESTED

15 Define the term *gene pool*.

Answer on p. 207

Hardy–Weinberg predictions

The Hardy–Weinberg principle predicts that allele frequencies *will not change* from generation to generation, as long as the following conditions are met:

- There is *no* selection taking place — all genotypes give individuals an equal chance of surviving and reproducing.
- There is no emigration or immigration, i.e. no individuals join or leave the population.
- The population is large. When the population is small, chance plays a large part in determining allele frequency. This is called **genetic drift**.
- There is no mating of individuals from different generations.
- There is no mutation, which could create new alleles.

A good working definition of evolution is 'when natural selection leads to change in allele frequency from one generation to the next'. However, as the above list shows, there are many other actors that can affect allele frequency.

Evolution

What is natural selection?

REVISED

Natural selection occurs when an individual has a genotype that gives it an advantage, which means that the individual passes on more of its alleles to the next generation. Individuals that pass on a lot of their alleles are said to be 'fit'. The frequencies of these beneficial alleles will therefore increase in the next generation.

Typical mistake

Don't use the term 'survival of the fittest' — it is a popular term that makes no sense to a proper biologist. In biology, fitness is a measure of reproductive success, so a 'fit' organism is one that passes on more of its alleles than other organisms. 'Survival of the best reproducers' is nonsense.

Types of selection

There are three types of natural selection: directional, stabilising and disruptive.

- **Directional selection** (Figure 10.8) occurs when individuals with an extreme phenotype have an advantage, such as the fastest, largest or most tolerant to cold. As a result, one phenotype becomes rare and an alternative becomes common.
- **Stabilising selection** (Figure 10.9) occurs when individuals with extremes of phenotype are at a disadvantage. In this case, individuals with intermediate phenotypes are more likely to pass on their alleles to their offspring. A good example of this is birth weight in mammals. Babies born too small have all sorts of problems, including susceptibility to cold. Particularly large babies can cause problems during birth.
- **Disruptive selection** (Figure 10.10) occurs when individuals at either extreme of phenotype have an advantage. A good example of this is seen in Darwin's finches, in the Galapagos islands. Imagine a population of finches that have a range of beak sizes; those with very small beaks have an advantage because they can get small insects out of small spaces such as cracks in tree bark, while those with very large beaks have the strength to crush seeds, or to catch larger prey. Both extremes have an advantage over those individuals with medium-sized beaks.

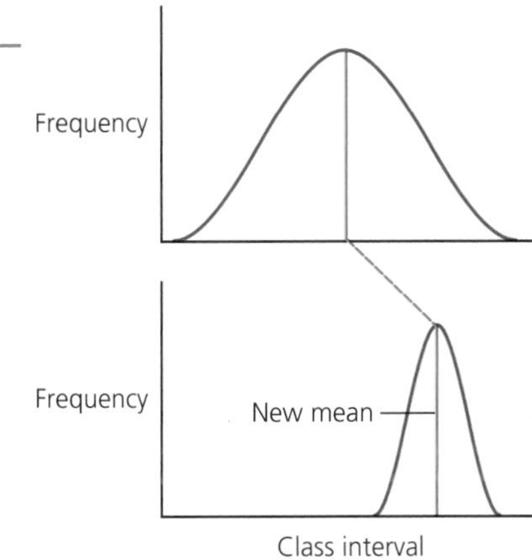

Figure 10.8 Directional selection

In Figure 10.8, the upper graph shows a variation in one factor. The mode (most frequent value) is marked in red. The graph represents the frequency distribution of this population before directional selection has occurred. The lower graph shows the same population after directional selection has occurred. As you can see, selection favours the individuals at one extreme, so in the following generations the mode shifts to the right.

In Figure 10.9, starting with the same graph as Figure 10.8, stabilising selection selects against both extremes, resulting in the same modal value but less variation.

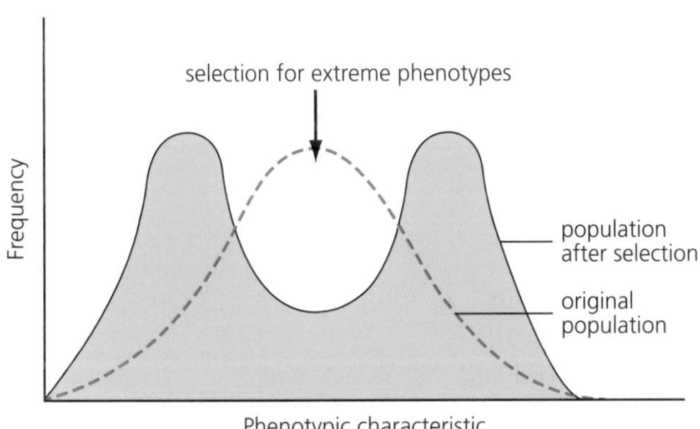

Figure 10.10 Disruptive selection

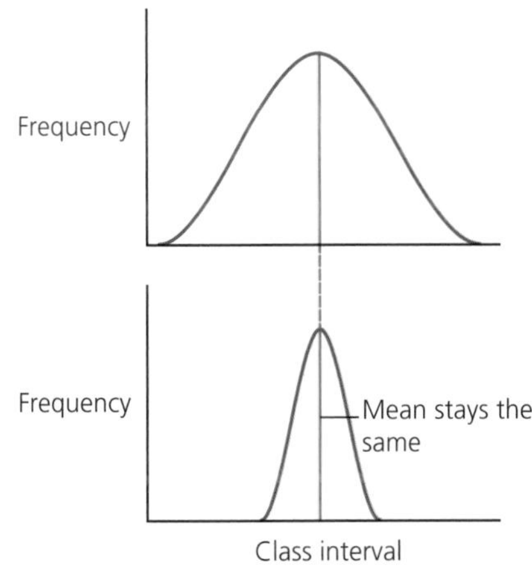

Figure 10.9 Stabilising selection

Speciation

How do new species develop?

REVISED

New species develop from a common ancestor. There are five basic steps involved:

1. The original population splits and the two or more populations become separated from each other so that they cannot interbreed.
2. There is variation within the population — a variety of different alleles producing many different genotypes.
3. Selection acts differently on the two populations.
4. Allele frequencies change in the two gene pools.
5. Over the generations, genetic differences accumulate so that even if the two populations were to mix they could not interbreed.

Allopatric and sympatric speciation

REVISED

New species evolve from common ancestors when natural selection acts differently on two or more populations within the same species, so the process of speciation cannot happen unless populations become **reproductively isolated**. The vital idea is that one population cannot interbreed with the others.

There are lots of different types of isolating mechanism, but the easiest one to understand is **geographic isolation**, which occurs when two populations cannot physically get in contact with each other. For example, if a road is built through a forest, this might separate two populations of a particular insect species.

There are two basic types of speciation:

- **Allopatric speciation** occurs when the two populations are physically separated. 'Allo' means apart. This type of speciation is easy to observe and there are many clear examples.
- **Sympatric speciation** occurs despite the fact that the two populations are together. 'Sym' means together. This type of speciation is very difficult to observe and some scientists dispute that it exists at all.

The importance of genetic drift

REVISED

When populations are small, *chance* plays a large part in determining allele frequency. Simple luck can determine how many alleles an individual passes on, and when a population is small, this can have a significant effect on allele frequencies. As a silly example, imagine a shipwreck in which just one lifeboat survives, containing just two families. They all have blond or ginger hair. They find themselves on a deserted island and survive, founding a new population. Purely by chance, the alleles for blonde and ginger hair have become frequent, at the expense of alleles for dark hair. In subsequent generations, the allele frequency will reflect that of the originals — this is the **founder effect**.

Genetic drift can be an important force in evolution but, unlike natural selection, it doesn't produce adaptations.

Exam practice

1 In a particular species of beetle, the allele for green wing-cases is dominant over the allele for black wing-cases. For example, in a population there are 400 individuals, of which 64 have black wing-cases.

(a) Calculate the percentage of beetles that are heterozygous. [4]

(b) The Hardy–Weinberg principle predicts that allele frequencies will not change from year to year, but scientists have found that the frequency of the black wing-cased beetles is decreasing. Suggest an explanation. [2]

2 Explain how two species can evolve from a common ancestor. [4]

Answers and quick quiz 10 online

ONLINE

Summary

By the end of this chapter you should be able to understand the following:

- An organism's phenotype is a result of the genotype and its interaction with the environment.
- Alleles are one or more alternative versions of the same gene. They may be dominant, recessive or codominant.
- Some genes are sex linked, in which case the XX female will have two copies whereas males have only one. This affects the pattern of inheritance.
- The Hardy–Weinberg principle allows allele and genotype frequencies to be calculated using the formula $p^2 + 2pq + q^2 = 1$.
- The Hardy–Weinberg principle predicts that allele frequencies will not change from one generation to the next, unless there is selection, migration, mutation or a small population.
- Natural selection occurs when some genotypes have greater reproductive success than others. This affects allele frequency within a gene pool. Selection can be directional, producing change by selecting for an extreme of phenotype, or stabilising, which selects against the extremes.
- New species arise as a result of reproductive isolation followed by different selection pressures on the different populations. Over the generations, genetic differences accumulate so the populations cannot interbreed.

11 Populations in ecosystems

Biotic and abiotic factors

A **population** is all the individuals of a particular species that can interbreed. A **community** is all the populations in the ecosystem, including plants, animals and decomposers (**saprobionts**). This section is about the ways in which organisms are affected by factors in their surroundings, which can be either biotic or abiotic.

The most common **biotic** (living) factors include:

- food supply
- **predation**
- competition for mates and nesting sites
- diseases such as bacteria, viruses, fungi and parasites

The **abiotic** (non-living) environment includes:

- temperature
- light
- inorganic nutrients such as nitrate and phosphate
- carbon dioxide and oxygen
- pH
- humidity

There are also many other factors that affect organisms, depending on their **habitat**.

In year 1 we looked at **diversity**. As a general rule:

- **Ecosystems** that have a **high diversity** have a favourable abiotic environment. For example, rainforests have plenty of sunlight, a favourable temperature, lots of water and a steady climate without unfavourable seasons. In these conditions, biotic factors tend to dominate an organism's life. There are thousands of different **niches** for different species.
- In contrast, in ecosystems with a **low diversity** such as deserts and Arctic regions, abiotic conditions dominate. Not many organisms can survive low temperature and/or a lack of available water. Sometimes there is no water available at all, and sometimes water is not available because it is frozen or too salty.

Saprobionts are microorganisms that break down the organic compounds in dead plants and animals and release carbon dioxide and simple inorganic ions.

Predation simply means one organism eating another. We generally think of predators as carnivores, such as cats, but plants are preyed on too. Herbivores prey on plants.

An **ecosystem** is a natural unit consisting of living and non-living components. Examples include a coral reef, a freshwater lake or a temperate forest.

A **niche** is a concept that describes an organism's role in the ecosystem. Each species occupies a particular niche because it is adapted to a particular set of biotic and abiotic factors. No two species occupy exactly the same niche.

Now test yourself

TESTED

1 Distinguish between the terms *population* and *community*.
2 Oak trees are a common woodland species. List *three* biotic and *three* abiotic factors that might affect an individual oak tree.

Answers on p. 207

Investigating populations

Practical techniques

REVISED

Investigating ecosystems involves **random** and **systematic sampling** in order to obtain **quantitative data**. You cannot possibly measure all that there is to measure, so the aim is to make sure that your sample is representative of the data as a whole. Data must be collected **randomly**, which means **without bias** (the conscious choice of the experimenter).

Quadrats

Quadrats are frames that are used for comparing two areas of land, such as north- and south-facing walls or mown and un-mown fields. The species **abundance** and **distribution** can be measured, but this method is usually limited to plants, lichens and slow-moving animals such as snails or limpets.

How to use quadrats:

1 Map out the areas you want to study. This can be done virtually or actually (with pins and string, for example, or with two tape measures set at right angles).
2 Number each square or work out a coordinate system (like a chess board, perhaps A5, B6 etc. — Figure 11.1).
3 Select numbers to sample. This must be done at random to avoid bias. Acceptable ways to do this include using a random-number generator, a telephone directory or a book of random numbers.
4 Take readings at the selected quadrats. What you actually do depends on the organisms in the sample. You could count the individuals to work out their frequency or estimate their percentage cover, or you could rate a particular species as abundant, common, frequent, occasional or rare (the ACFOR scale).

Typical mistake

Referring to quadrants rather than quadrats.

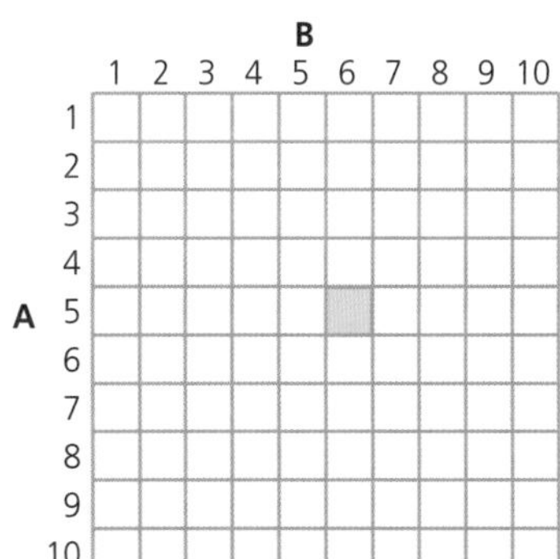

Figure 11.1 A grid with a quadrat positioned at A5, B6

Transects

Transects are lines that are used to show change from one area to another. They can be used on their own or with quadrats. Situations where you might use transects include rocky shores, sand dunes and rivers. Different types of transect include:

- point transects — simply record the species present at points along the line
- belt transects — place a quadrat continuously alone the line and record the organisms in it
- interrupted belt transect — for longer surveys, a quadrat can be placed at regular intervals, say every 5 m or 10 m (Figure 11.2)

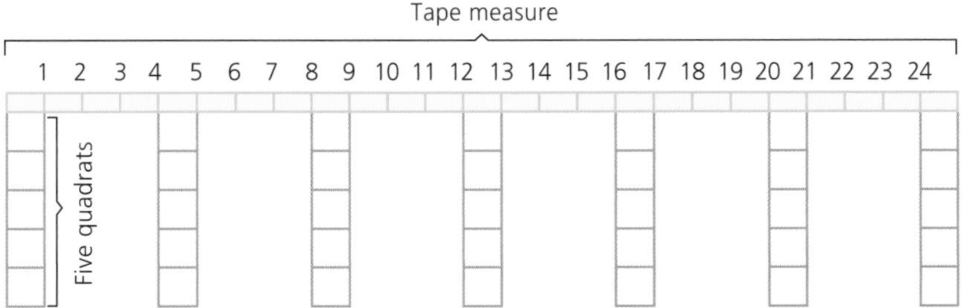

Figure 11.2 An interrupted belt transect; one or more quadrats can be used at every interval

Exam tip

The use of quadrats to study two different areas usually involves **random** sampling, while the use of transects involves **systematic** sampling.

Now test yourself

TESTED

3 Which type of sampling technique would be most suitable for investigating each of the following?
   (a) the growth of moss and lichens on north- and south-facing tree trunks
   (b) the distribution of organisms down a rocky shoreline
   (c) the distribution of organisms across a sand dune from the sea

Answer on p. 207

Mark-release-recapture

Animal populations can be measured using **mark-release-recapture**. To estimate the population of animal species, the process is as follows:

1 Catch the animals. Pitfall traps work for small species like beetles, whereas Longworth traps are good for small mammals.
2 Mark them in a way that does not harm them, restrict their mobility or make them more visible to predators. A spot of paint works for snails, whereas a small patch of fur can be clipped when marking mammals.
3 Let them go and give them enough time to mix with the rest of the population.
4 Set traps again — keeping all methods and timings the same — and note how many are marked.

Exam tip

There are many different ways of trapping animals and you are not expected to know all of them. You can suggest improvisations in any given situation, such as pooters (where small invertebrates are sucked up into a container), nets for butterflies and grasshoppers and kick samples for river or stream organisms. You can even beat tree branches with sticks and collect the insects that fall in a tray, sheet or an old umbrella.

The population can be estimated using the formula:

$$\text{population} = \frac{\text{number in first sample} \times \text{number in second sample}}{\text{number marked in second sample}}$$

For example, in an investigation into a population of wood mice the following data were obtained:

First sample = 75 mice captured, marked and released

Second sample = 81 mice captured, of which 8 were marked

Therefore, the population is calculated as:

$$\frac{75 \times 81}{8} = 759 \text{ (rounded down)}$$

Now test yourself

TESTED

4 To estimate the population of periwinkles in a harbour, 200 individuals were collected and marked with a spot of paint. The next day another 200 were collected, of which 23 were already marked. Estimate the population.

Answer on p. 207

How ecosystems develop

From pioneer species to climax community

REVISED

How do we go from bare rock, sand or sterile soil to a fully developed, mature forest? The process involves **colonisation** and **succession**, as shown in Figure 11.3.

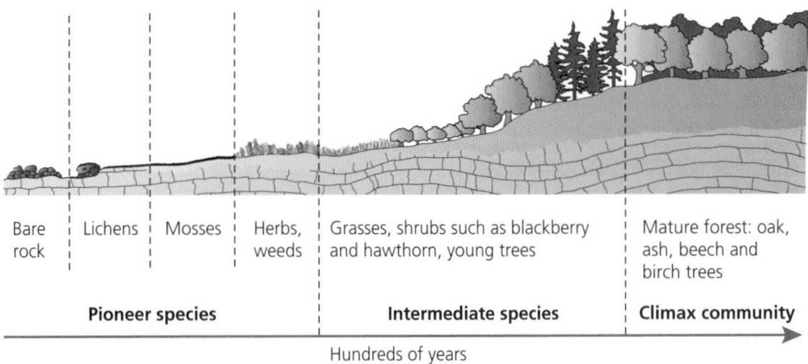

Figure 11.3 Succession from bare rock

In the beginning, there is a harsh, **abiotic environment**, such as sand or bare rock. On land, nutrients and water are in short supply. Only a few specialist organisms can colonise: the **pioneer species**. When the pioneer species have become established, they make the environment more favourable and **less hostile**. This allows other organisms to survive, which then succeed the colonisers by out-competing them. Succession continues as the conditions become increasingly favourable, with different species dominating at each stage. Succession stops when a **climax community** is established. This is defined as a stable co-existence of dominant species.

For example, starting with bare rock, the pioneer species are usually lichens — fungi that contain algae. This is a **mutualism** — an association in which both benefit. The fungi provide carbon dioxide and nitrogen as well as protecting the algae from drying out. In turn, the algae photosynthesise and provide organic molecules such as sugars. Each provides what the other needs, so they can survive in harsh places.

Once the lichens are established, they change the conditions so that simple plants such as mosses can get a hold. The mosses hold moisture and trap soil particles so that more complex plants such as ferns can become established. More and more soil accumulates under and between the plants, so that moisture and humus build up. Succession continues until the plants with woody stems dominate — the shrubs and trees. These large, tall plants out-compete the others for light and can survive from one year to the next, growing larger and larger.

In the UK, the climax community is a deciduous forest — a stable population of dominant tree species that shed their leaves each year. Much of the UK was covered in this type of forest, until humans cut it down. Globally, the climax community that develops depends on the climate. Examples include tropical rainforest, temperate coniferous forest and tundra.

Grazing

REVISED

Succession cannot take place if there is a lot of grazing. Sheep, rabbits, cows and deer can all prevent the ecosystem from developing beyond grassland because they remove the growing tips from many of the plant species, including tree saplings. Grass has its growing point at the bottom of the stem, so if the tip is removed it still keeps growing.

Now test yourself

TESTED

5 What is colonisation?
6 Why does succession happen?
7 What is a climax community?
8 What is humus?

Answers on pp. 207–8

Management of succession

REVISED

There are situations where the management of succession brings benefits, such as:

- maximising the growth of trees
- maximising habitat **diversity**
- maintaining the look of the countryside

For example:

- The regular burning of grouse moors encourages new growth of heather shoots, which is a perfect habitat for grouse.
- Coppicing (cutting down some large trees to keep the succession at an intermediate stage) can stimulate new tree growth, creating more niches for endangered species.
- In the Lake District, sheep farming is encouraged because other practices would dramatically change the landscape — the look of the countryside is important for tourism.

Now test yourself

TESTED

9 What is meant by the term *net productivity*?
10 Suggest suitable units for measuring net productivity of a woodland.
11 The net productivity of an ecosystem changes as it develops. Productivity tends to be low at first, then relatively high, before falling once again as the ecosystem stabilises. Suggest reasons for each of the three changes in productivity.

Answers on p. 208

Summary

By the end of this chapter you should be able to understand the following:

- In a habitat, a population consists of all the individuals of one species, whereas a community consists of all the individuals of all the species.
- Each species occupies a particular niche, governed by its adaptation to both biotic and abiotic conditions.
- Quadrats and transects are used for random sampling of data. Quadrats compare different areas of land, whereas transects show gradual change from one area to another.
- Animal population size can be estimated using the mark-release-recapture method.
- Ecosystems develop by a process of colonisation and succession.
- At each stage in the succession, the dominant species change the abiotic environment so that it is more favourable for other species.
- Conservation of habitats often involves the management of succession, so that diversity is maintained.
- Ecosystems develop by a process of colonisation by pioneer species, followed by succession that ends with a climax.
- At each stage in succession, certain key species change the environment so that it becomes more suitable for other species.
- The changes in the abiotic environment result in a less hostile environment and changing diversity.
- Conservation of habitats frequently involves management of succession.

Exam practice

1 In an investigation, the population density of plants in a regularly cut lawn was compared with that in a lawn that was cut only occasionally. The results are shown in the table.

	Mean population density/number of plants per m^2		Result of statistical test
Species	**Regularly mown lawn**	**Occasionally mown lawn**	**Value of *p***
Clover	35.0	18.1	< 0.02
Dandelion	10.8	3.4	< 0.05
Ragwort	1.2	8.7	< 0.01
Buttercup	1.3	10.1	< 0.01
Ribwort plantain	4.3	2.8	> 0.5

(a) Describe how you could use quadrats to find the mean population density of ragwort plants on a lawn. [4]
(b) Give the null hypothesis for the statistical test on the population density of clover. [1]
(c) Use the words *probability* and *change* to explain what is meant when $p < 0.05$. [2]
(d) What conclusions can be drawn from the results of this investigation? [3]

2 In April greenfly hatch from eggs that were laid on a rose bush the previous year. The greenfly begin to reproduce asexually. All the offspring are wingless females, born with young greenfly already developing inside them. After a few months there is a large population on the plant. At this point the aphids begin to give birth to winged males and females that can fly away, mate and lay eggs on new bushes.
(a) Suggest why it is an advantage to reproduce asexually in spring. [3]
(b) Competition is a *density-dependent factor*. Explain what this term means. [1]
(c) List *two* limiting factors that could slow down population growth. [1]
(d) Suggest the advantages of the following:
(i) wings [1]
(ii) sexual reproduction [1]
(iii) laying eggs [1]

3 A student wanted to sample the diversity of plant species in a meadow. She needed to know how many quadrats to use in the investigation, so she added up the number of species identified. Explain how she could use the following graph to decide how many quadrats to use. Use point X to help you. [2]

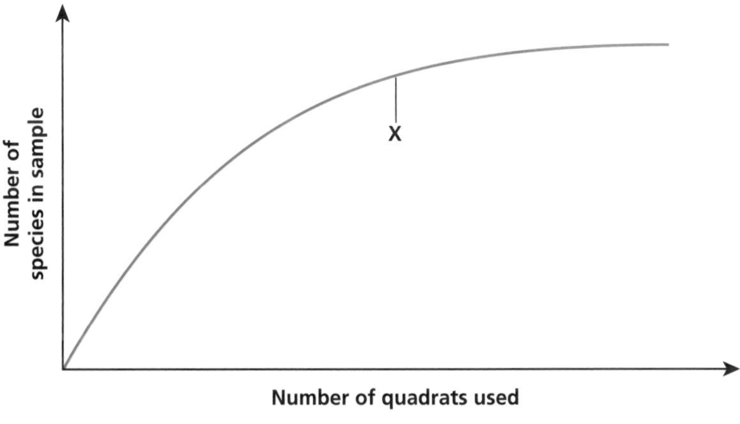

4 (a) Define the term *biodiversity*. [2]
(b) Biodiversity generally increases as an ecosystem develops. Explain why. [3]
(c) Coppicing is a practice that cuts down trees in a systematic manner so that they cannot grow to their full height, but which encourages growth of new shoots and smaller shrubs. Explain the conservation benefits of coppicing. [2]

Answers and quick quiz 11 online

ONLINE

12 The control of gene expression

Mutation: alteration of the sequence of bases in DNA can alter the structure of proteins

Gene mutations

REVISED

Gene mutations are changes in the DNA of a cell that result in a changed base sequence. The commonest cause of mutation is a mistake in **DNA replication**. These mistakes are relatively rare — one estimate puts the mutation rate as one mistake in every 10^{10} times a base is inserted, but given the number of cells, and the number of bases in each cell, they are a regular occurrence.

Mutation can occur in two places:

- in **somatic** (body) cells. These mutations accumulate during the life of an organism — and can result in cancer — but are not passed on to the offspring.
- in **gametes** (sex cells). These are known as **germline** mutations. These mutations alter the genotype of the individual and will be present in every cell.

Mutations might or might not affect the organism. Mutations might *not* affect the organism because:

- some take place in the non-coding DNA between genes
- some take place in the introns (non-coding sequences) within genes
- some will still code for the same amino acid — the genetic code is **degenerate**. For example, if the codon GUU mutates to GUC, GUA or GUG, it will still code for the amino acid valine
- some will cause a change in the amino acid sequence, but will not significantly change the tertiary structure. The protein is still the right shape and will still function in the organism

Therefore, the only gene mutations that affect organisms are the ones that bring about significant changes in the structure of the protein. There are several different types of mutation, which alter the base sequence in different ways:

- **Deletion**, in which one base pair is lost. As a consequence, all the bases move along in one direction and therefore many codons are changed. This is known as a **frame shift**.
- **Addition**, in which one base is added and all the bases after the mutation are pushed along by one. This also results in a frame shift.
- **Substitution**, in which one base pair is substituted for another. This is also called a **point mutation**. Only one codon is changed, which might or might not change the amino acid in the protein. Remember that the genetic code is degenerate, meaning that several different codons can code for the same amino acid.

- **Inversion**, in which a base sequence or whole gene is taken out, rotated 180° and inserted backwards. Again, the protein that results from the translation of an inversion will be different and probably not functional.
- **Duplication**, in which a sequence of bases or a whole gene is inserted twice, or multiple times. This is important in evolution because one of the genes can mutate, and take on a new function, while the original continues to work as normal.
- **Translocation**, in which a sequence of bases or a whole gene is taken out and inserted at a different position on the chromosome, or even on a different chromosome. This usually results in a non-functional protein.

Table 12.1 illustrates the different types of mutation and Table 12.2 shows the effects of these mutations. The changed bases are shown in red.

Table 12.1

Original sequence	CATAGGTGAC
Deletion	If the 4th base along, A is deleted, the result is CATGGTGAC because there is a frame shift to the left
Addition	CATCAGGTGAC If the base C is added at the position shown, the result is a frame shift to the right
Substitution	CATAGGTGAC If the underlined base A is substituted for G, only one codon is changed: CATGGGTGAC
Inversion	CATAGGTGAC If the underlined sequence is inverted, the result is CATACAGTGG which, in this case, changes at least two codons
Duplication	CATAGGTGAC If the underlined sequence is duplicated once, the result is CATAGGTAGGTGAC, which changes three codons
Translocation	CATAGGTGAC If the sequence TAG is translocated, the result could be CAGTGACTAG, which changes both the old location and the new one

Table 12.2 **The effect of three different types of mutation. The first two rows show the original base sequence. The three examples beneath show the effects of deletion, substitution and insertion mutations. Note that deletion and insertion mutations both cause a frame shift, causing a much greater change in the amino acid sequence than the substitution**

Original base sequence on mRNA	AGA	UAC	GCA	CAC	AUG	CGC
Encoded sequence of amino acids	Arg	Tyr	Ala	His	Met	Arg
mRNA base sequence after deletion of the last base on the first codon (frame shift)	AGU	ACG	CAC	ACA	UGC	GCX
Encoded sequence of amino acids	Ser	Thr	His	Thr	Cys	Ala
mRNA base sequence after base substitution	AGC	UAC	GCA	CAC	AUG	CGC
Encoded sequence of amino acids	Ser	Tyr	Ala	His	Met	Arg
mRNA base sequence after base addition (frame shift)	AGG	AUA	CGC	ACA	CAU	GCG
Encoded sequence of amino acids	Arg	Ile	Arg	Thr	His	Ala

With the exception of substitution, all of the mutations described above have the potential to seriously disrupt the polypeptide and therefore the protein that is being produced.

Exam tip

You cannot appreciate the effects of mutation without knowing the basics of protein synthesis, which you covered in year 1. You will remember that a gene is a sequence of bases that codes for the amino acid sequence of a polypeptide/protein. The process of transcription makes mRNA — a mobile copy of a gene — and then (once the introns have been spliced out) the process of translation uses the mRNA to make the protein.

Exam tip

Don't make the mistake of thinking that all mutations are harmful. Many have no effect and a vital few are beneficial.

Now test yourself

TESTED

1 Explain how the gene for a particular enzyme may mutate but still produce a functional enzyme.
2 Suggest why germline mutations are more common in males than in females.

Answers on p. 208

Gene mutations occur randomly, at a slow and steady rate, and any part of the DNA can mutate — the genes or the DNA between them. Most of these mistakes are spotted and corrected by a 'proofreading' mechanism within the cell. However, this rate can be increased by **mutagenic agents**, which include:

- some chemicals, including benzene, mustard gas and bromine/bromine compounds
- ionising radiation (gamma and X-rays)
- ultraviolet light
- biological agents such as some viruses and bacteria

Gene expression is controlled by a number of features

Stem cells

REVISED

Multicellular organisms have many different types of specialised cell. The human body, for example, has over 200 different types. The key to cell specialisation is the **selective activation of genes**. In muscle cells, for example, the genes that code for the contractile proteins actin and myosin need to be activated. There is a huge amount of DNA in every cell and the vast majority of it is never translated into proteins. In any particular cell, only specific genes are active.

Stem cells are unspecialised cells with two key properties:

- **self-renewal** — the ability to divide again and again
- **potency** — the ability to become specialised. Usually, this is a one-way process; as the cells specialise, they lose their potency.

The cells of the early embryo are **totipotent**. This is illustrated by the fact that an embryo can be pulled apart and each half, fragment, or even each individual cell has the power to develop into a whole new individual. This is one method of **cloning**.

Now test yourself

TESTED

3 Explain what is meant by the term *clone*.

Answer on p. 208

There is a scale of potency from the most potent (totipotent) to the least potent (unipotent).

Totipotent cells occur only for a limited time in very early mammalian embryos. They have the power to differentiate into any of the specialised cells that make up the body, and the cells that support the embryo, such as the placenta. Therefore, these cells have the power to make whole organisms.

Pluripotent cells have the power to develop into any of the specialised cells of the body but not the placenta, so they don't have the power to generate a whole new individual. After a few days of development, a human embryo develops into a hollow ball called a **blastocyst** (Figure 12.1) and the **inner cell mass** contains pluripotent stem cells.

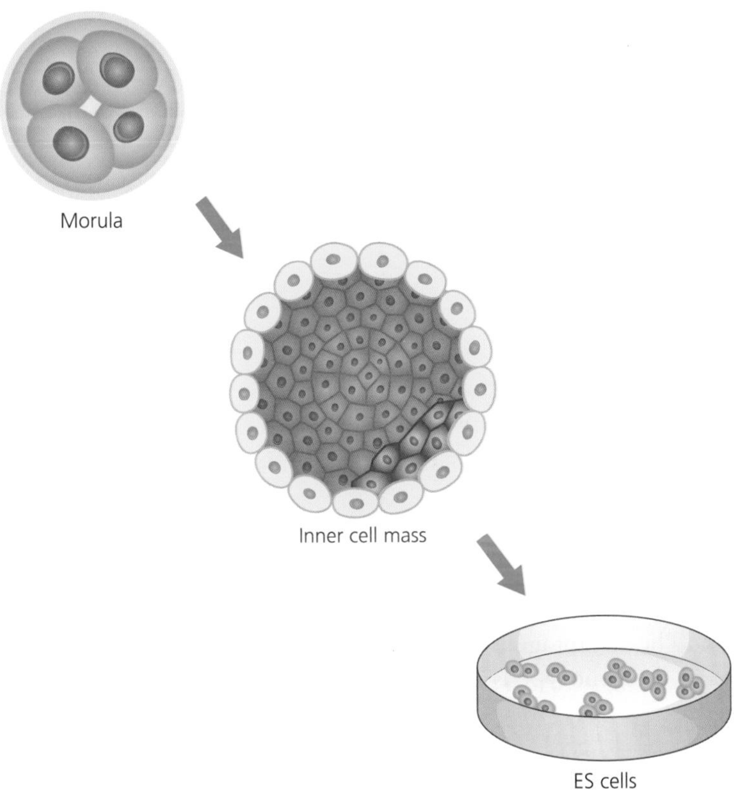

Figure 12.1 **The early embryo, about 7 days old, is called a blastocyst. Pregnancy begins when the blastocyst implants in the uterus wall. The outer cells form the placenta while the inner cell mass will develop into the foetus itself**

Multipotent cells have the ability to differentiate into a limited number of cell types. In the adult there are some tissues, notably brain and bone marrow, that contain multipotent stem cells. In the bone marrow, **haematopoietic** cells can differentiate into red cells and the various different types of white cell.

Unipotent cells have the ability to differentiate into just one cells type. In the heart, for example, **cardiomycetes** are cells that differentiate into new heart muscle cells.

There is great potential for using stem cells in medicine. Bone marrow transplants and the generation of new skin for burns victims are long-established treatments, because they don't involve reprogramming the stem cells. A lot of research is being done at the moment with the aim to be able to reliably persuade stem cells to differentiate into a particular cell type. If and when we can achieve this, the potential applications are enormous. They include:

- replacement β cells to treat type 1 diabetics
- repair of damaged heart muscle
- repair of brain cells for Parkinson's disease
- retinal repair for eye defects

Evaluating stem cell research

Balanced against the great potential of stem cells are significant scientific and ethical problems. The scientific problems are simply that we cannot reliably reprogramme stem cells. Even if they reliably differentiate and function in vitro, there is no guarantee that they will do the same in the body. There is also the possibility that the stem cells will begin to multiply out of control, and cause tumours.

The key ethical issue is the use of embryonic stem cells. There is no shortage of human embryos because infertility treatment creates more than are needed. Some people say that the embryo is a human life with the potential to grow into a new individual and therefore should be given rights. Others argue that an embryo is a tiny ball of cells without a name or a nervous system, incapable of feeling pain, which would otherwise simply be destroyed.

Exam tip

The command word *evaluate* means 'look at both sides'.

Stem cells in plants

In mammals, only early embryonic stem cells are totipotent, but most plant cells remain totipotent in adulthood. This allows whole plants to be **cloned** from a simple tissue sample or cutting. For example, a piece of cabbage leaf, if placed in sterile conditions, can develop into a whole new cabbage plant in a short period of time. This is useful because:

- plants with desirable characteristics can be reliably cloned
- some valuable crops — orchids, for example — can be grown quickly with a minimum of space and time

Cloning is an example of asexual reproduction. The advantage is that there is no mixing of genetic information and so the genotype is always preserved unless there is a mutation.

The control of gene expression

REVISED

Gene expression involves the following flow of genetic information: DNA → mRNA → polypeptide. It is possible to control the expression of genes at any stage in the process, for example by:

- preventing transcription
- interfering with the splicing of the introns
- destroying the mRNA
- interfering with translation (e.g. by blocking the ribosome)
- preventing a polypeptide from being turned into a functional protein

The control of gene expression has great potential in medicine, particularly in the treatment of cancer. The next section features some of the main areas of research.

Now test yourself

TESTED

4 Explain what will happen if a piece of mRNA is translated without having the introns removed.

Answer on p. 208

Regulation of transcription and translation

Transcription of target genes

REVISED

Transcription of a gene starts when the RNA polymerase enzyme binds to the start of the gene and begins to make mRNA (Figure 12.2). However, this can only happen when all the **specific transcriptional factors** are in place (Figure 12.3). These factors combine to form the **transcription initiation complex (TIC)**. Some of the factors are present in the **cytoplasm**, moving into the **nucleus** when needed, and some of them come from outside the cell. Some hormones act as transcriptional factors.

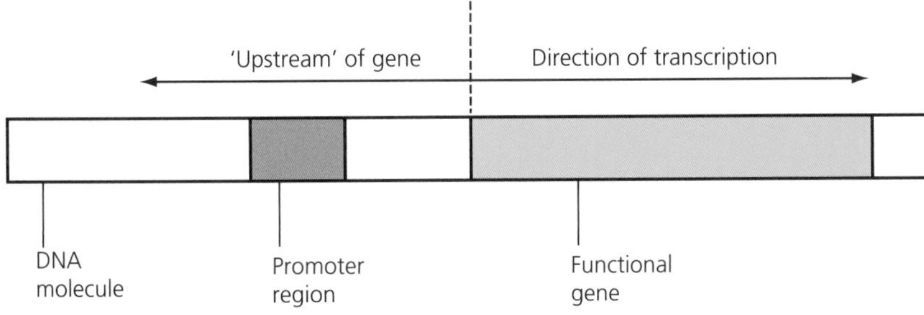

Figure 12.2 In eukaryotes, all genes have a promoter region that is described as being upstream of the gene

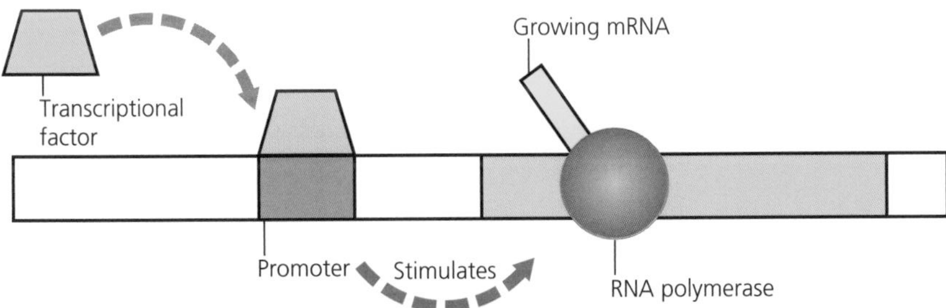

Figure 12.3 The functional gene contains the code for making the polypeptide, but it won't be transcribed unless the correct transcriptional factors attach to the promoter

Oestrogen and gene transcription

Oestrogen is a steroid hormone (page XX). As steroids are lipids, they pass through the cell membrane. Oestrogen passes into the nucleus of target cells, where it combines with an oestrogen receptor called ERα (estrogen factor alpha — using the American spelling of oestrogen). This forms an active ERα complex that passes into the nucleus and acts as a transcriptional factor for many different genes.

Epigenetic control of gene expression in eukaryotes

REVISED

Epigenetics is the study of the way in which environmental factors can affect gene expression. The genes themselves don't change, but the potential for them to be expressed does. Vitally, epigenetic changes can be inherited.

There are two epigenetic mechanisms that switch genes off:

- increased methylation of DNA
- decreased acetylation of associated histones

Methylation is simply the addition of the methyl ($-CH_3$) group to the DNA, and specifically to the base cytosine. A gene with methylated cytosines is often said to be 'capped' and is less likely to be expressed.

A gene cannot be transcribed unless it has been unwound from its histone proteins. Acetylation is the addition of an acetyl group ($COCH_3$) to the amino acid lysine on the histone, which reduces the attraction between the DNA and the histone. As a result, the DNA tends to detach from the histone, thus allowing transcription. Decreased acetylation therefore increases the attraction between DNA and histone, making the genes more tightly wound and unable to be transcribed (Figures 12.4 and 12.5).

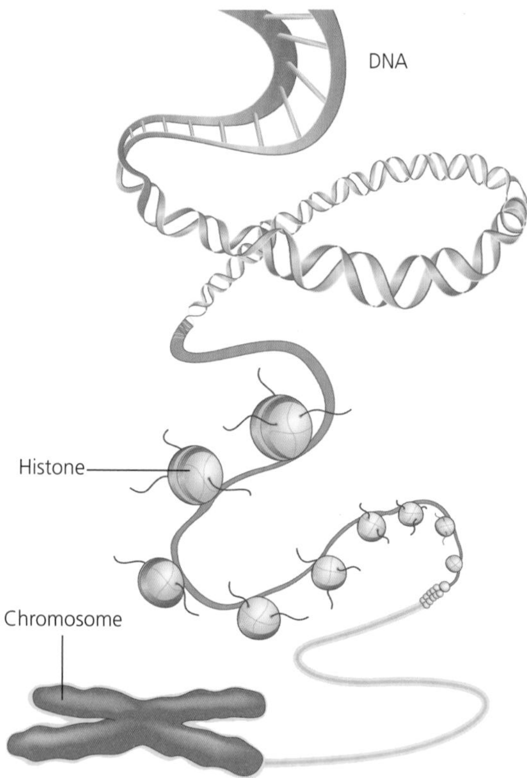

Figure 12.4 DNA molecules are incredibly long, and are organised by winding round proteins called histones. DNA that is bound to a histone cannot be transcribed, in the same way that you cannot read a rolled-up newspaper

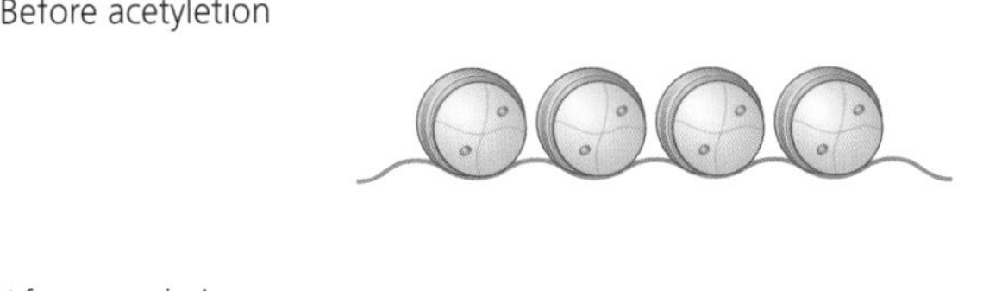

Figure 12.5 When histones are acetylated, the DNA becomes less tightly attached. This allows the genes to be transcribed

Epigenetics and cancer

It seems likely that epigenetic changes can play an important role in the development of some types of cancer. Epigenetic changes that increase the expression of an oncogene, or that silence a tumour suppressor gene, can lead to tumour development.

Small interfering RNA

REVISED

Small interfering RNA (siRNA) is a short, double-stranded molecule of RNA (Figure 12.6) that interferes with the expression of a specific gene by breaking down the mRNA and therefore preventing translation.

One molecule of siRNA combines with several proteins to form an RNA–protein complex. One of the strands of the siRNA comes away and the other is used by the RNA–protein complex to seek out and bind to the mRNA that needs silencing. It is another example of complementary base pairing. The mRNA is cut by the proteins and therefore cannot be translated.

It is thought that siRNA evolved as a defence against viral attack and it has great potential in medicine because it can prevent the expression of harmful genes. It is one example of the molecules used in the process of RNAi (RNA interference).

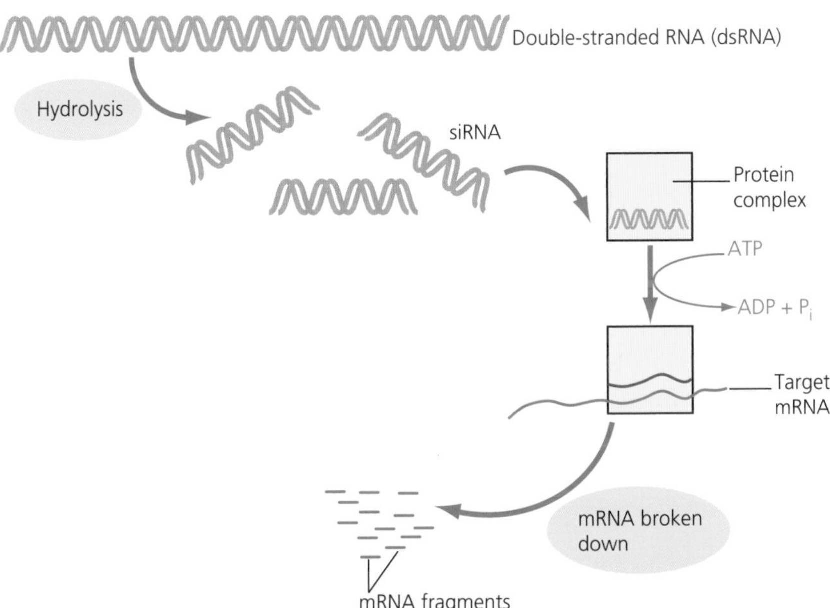

Figure 12.6 Small interfering RNA (siRNA) is a short, double-stranded molecule of RNA

Exam tip

It is easy to go into too much detail about the way siRNA works. Exam questions will not ask for details beyond the basic mechanism and its potential. Most questions will use siRNA to test your understanding of protein synthesis.

Gene expression and cancer

Cell division and tumours

REVISED

The control of **cell division** is vital. Cells should only divide when needed, so that body tissues replace themselves at the correct rate. If a cell's control system breaks down and cell division (mitosis) becomes too fast, a **tumour** can result.

The rate of cell division (mitosis) is controlled by genes called **proto-oncogenes**. If these genes mutate into **oncogenes**, cells may start to divide too quickly. There is a back-up system in the form of **tumour suppressor genes**, which prevent rapid cell division or cause the death of the cell if the damage cannot be repaired.

Generally, there are two types of tumour: **benign** and **malignant**. If cell division occurs at the centre of the tumour and does not spread, it is benign. If dividing cells are at the edge of the tumour and likely to break off and set up secondary tumours, it is malignant — that is cancer.

We accumulate these mutations throughout our lives and if the proto-oncogenes mutate and the tumour suppressor genes mutate, a tumour can result (Figure 12.7). There is another back-up system: the immune system. White cells can detect abnormal cells in the same way that they detect pathogens, so that small tumours are destroyed.

Exam tip

Think of a proto-oncogene as the accelerator on a car, and the tumour-suppressor gene as the brakes. If the accelerator jams on, and both the front and rear brakes fail, you're in trouble.

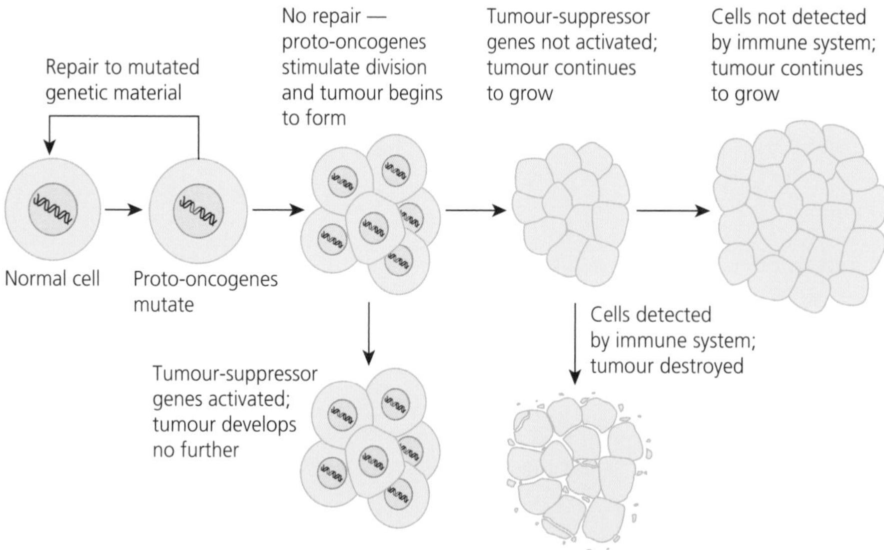

Figure 12.7 The development of tumours

Using genome projects

REVISED

The **genome** is all the genetic material in a single cell from an organism. In humans, we know it consists of 23 chromosomes, over 3 billion base pairs and our current best estimate is that there are about 21 000 genes. Incredibly, most body cells contain the entire genome twice, because they are diploid.

Typical mistake

Thinking that once we know an individual's genome, we can predict their future health with a degree of certainty. Lifestyle and chance have a great deal to do with it.

Now test yourself

TESTED

5 Name two types of cell in the human body that do not contain two copies of the genome.

Answer on p. 208

The human genome

REVISED

The ultimate goal is to understand the human genome and that of many other species. The aim is to find out where all the genes are, what they do and how they interact to build and maintain an organism. This is a fast-moving area of science and the technology advances on a daily basis.

The announcement that the entire human genome had been sequenced came in 2001, having started in 1990. The human genome and that of simple organisms such as yeast and *E. coli* were sequenced using slow and labour-intensive techniques, known as **first-generation** sequencing. There are now machines — known as **third-generation sequencers** — that can sequence an individual human's entire genome in a few hours, and we are constantly refining technologies that allow the machines to work faster. The sequencing is also getting cheaper — the sequencing cost per base is becoming significantly less.

There are many benefits to understanding the genome. Medics want to know what goes wrong in cases of genetic disease. If we can sequence an individual's genome we can tailor treatment to their particular genotype rather than treating people as though they were all the same. For example, some people will respond well to a particular drug, while it will have little effect on others.

An interesting consequence of our study of the genome is the **proteome** — the range of proteins that can be made. You will remember from year 1 that basically one gene makes one protein, but it has been found the human proteome consist of many more proteins than there are genes. One estimate puts the human proteome at around 100 000 different proteins — about five times the number of genes.

Exam tip

If we can map out an individual's genome at birth, this raises many ethical questions. If an individual is prone to particular genetic diseases or conditions, who has the right to that information? Is it ethical to withhold life insurance or a mortgage from an individual who might have a short life expectancy? Is it ethical for that individual to take out life insurance or a mortgage, knowing how long they have to live?

Exam practice

1 The following base sequence is taken from within a gene:

AAGGCTCCATTG

(a) What is the maximum number of amino acids that could be coded for by this sequence? [1]

(b) What is the minimum number of amino acids that could be coded for by this sequence? Explain your answer. [2]

(c) Write down the corresponding mRNA sequence that would result from transcription. [1]

(d) In a mutation, the red letter A was substituted.

(i) How many triplets would be changed? [1]

(ii) Give *two* reasons why this mutation might not affect the organism. [2]

(e) The whole gene was analysed and the percentage of bases is shown in the following table. Fill in the missing value. [1]

Base	A	T	G	C
Percentage in gene	20	29	22	

2 Explain how oestrogen affects gene expression in a target cell. [5]

Answers and quick quiz 12 online

ONLINE

Summary

By the end of this chapter you should be able to understand:

- The structure and function of mRNA and tRNA.
- That the genetic code consists of base triplets, which code for specific amino acids.
- That the genetic code is universal, non-overlapping and degenerate.
- The events of transcription as the production of mRNA from DNA and the removal of introns.
- The events of translation.
- That gene mutations are changes in the base sequence that result from faults in DNA replication.
- The role of mutations in the development of tumours.
- The difference between totipotent and multipotent stem cells.
- The potential uses and ethical issues relating to stem cells.
- The potential of stem cells in plants.
- The effect of oestrogen on gene transcription.
- The role of siRNA in blocking the expression of a specific gene.

13 DNA technology

Recombinant DNA technology

Gene technology, or **genetic engineering**, involves the manipulation of the DNA molecule. The key processes include:

- finding genes
- finding out their base sequence
- making artificial copies
- cloning them
- putting them into different cells and organisms so that they are expressed

The genetic code is described as being **universal**, which means that the same codons are transcribed and translated into the same amino acids in all organisms. This means that a gene from one species can be put into another, where it can be used to make exactly the same protein as it did in the donor species.

Where do we get DNA fragments from?

REVISED

DNA molecules are huge, so genetic engineering usually involves working with fragments of DNA, such as individual genes that make useful proteins. There are three ways to make DNA fragments:

- Cut specific pieces of DNA out of the genome using restriction enzymes.
- Work backwards from mRNA, using reverse transcriptase to make **complementary DNA (cDNA)**.
- Work backwards from the protein. If you know the amino acid sequence of a protein, there are 'gene machines' that will make a piece of DNA that will encode that protein.

Exam tip

Restriction enzymes come from bacteria, where they are weapons against viral infection. Their full name, **restriction endonuclease**, refers to the fact that they *restrict* viral growth by cutting *within* the *nucleic acid*.

Table 13.1 The genetic engineering toolkit

Tool	Job
Restriction endonuclease	An enzyme that cuts DNA at specific **recognition sites**; usually produces sticky ends (Figure 13.1) rather than clean cuts
Ligase	An enzyme that joins two pieces of DNA, such as complementary **sticky ends**, to form recombinant DNA
Reverse transcriptase	An enzyme that makes a DNA molecule from the corresponding mRNA — transcription in reverse; found in **retroviruses** such as HIV; DNA made this way is called **complementary DNA** (cDNA)
DNA or gene probes	Radioactive or fluorescently labelled fragments of DNA that seek out and bind to a target sequence; used to detect the presence of particular genes, for example in medical diagnosis
Plasmids	Small, circular pieces of DNA found in bacteria; very useful as vectors for putting genes into bacteria

Sticky ends are staggered cuts, revealing unpaired base sequences at the ends of a piece of DNA.

Exam tip

The term 'gene machine' is a general term for several different technologies that can make artificial genes using the base sequence on mRNA or the amino acid sequence of a protein. All DNA made in this way is known as **complementary DNA**, or **cDNA**.

Exam tip

The exam may ask you about the basics of genetic engineering, but you will not be expected to know about the very latest developments.

Exam tip

When one piece of double-stranded DNA is joined to another — usually from a different species — the result is **recombinant DNA**. In contrast, when a piece of DNA contains two strands from different sources, that's called **hybrid DNA**. When a probe or a primer finds a complementary sequence, it hybridises.

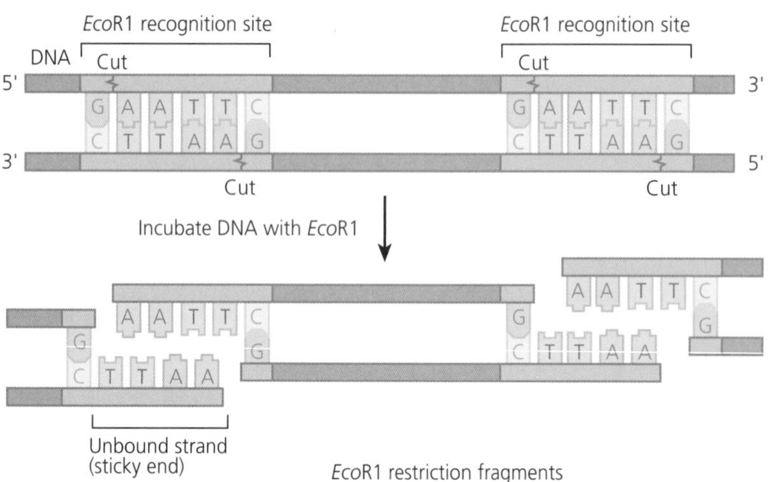

Figure 13.1 The restriction enzyme *Eco*R1 cuts DNA at the specific recognition sequence GAATTC. This leaves two sticky ends that can be joined to any other piece of DNA that has been cut by the same enzyme

In English, a palindrome is a word or phrase that is spelt the same forwards as backwards, such as *gnu dung*. It is not the same in biology. A **palindromic** DNA base sequence is the same on the other strand, reading in the other direction. For example, the sequence GCTAGC would read CGATCG on the other strand, which is the identical sequence backwards.

In vivo and in vitro techniques for amplifying DNA

REVISED

In vivo means 'in life' and refers to processes carried out in living cells and organisms. **In vitro** means 'in glass' and refers to processes carried out in a test tube or a Petri dish rather than in living material.

In vivo cloning involves putting DNA into a living cell so that the gene is copied each time the cell divides. In addition, the gene can be expressed because the whole cell contains ribosomes and all the other components needed to make proteins. In vitro cloning involves making copies of DNA using the **polymerase chain reaction (PCR)**.

The **polymerase chain reaction (PCR)** is a process used by biologists to make large amounts of identical DNA from very small samples.

The polymerase chain reaction (PCR)

REVISED

The PCR is gene cloning in a test tube. It is an effective and quick way of amplifying small samples of DNA, so there are multiple copies of useful genes or enough DNA for a DNA profile, for example.

The PCR just needs four simple ingredients:

- the DNA template to be copied
- DNA polymerase to copy the DNA — a thermostable enzyme is needed

- a supply of nucleotides
- primers — short sections of single-stranded DNA that allow the enzyme to attach and start copying

The thermostable enzyme commonly used is **Taq polymerase**. It is extracted from *Thermus aquaticus*, a bacterium found in hot volcanic springs. The enzyme is not denatured even at 94°C, so it can function at all the different temperatures encountered in the PCR cycle.

The three key steps in each PCR cycle (Figure 13.2) are as follows:

1 Denaturation — heat the mixture to 94°C to denature the DNA; the hydrogen bonds break and the two strands separate, yielding single-stranded DNA molecules.
2 Annealing — cool the mixture to 50–60°C to anneal (attach) the primers to the single-stranded DNA template.
3 Extension — heat the mixture to 74°C; the Taq polymerase moves along the DNA, catalysing the addition of complementary nucleotides.

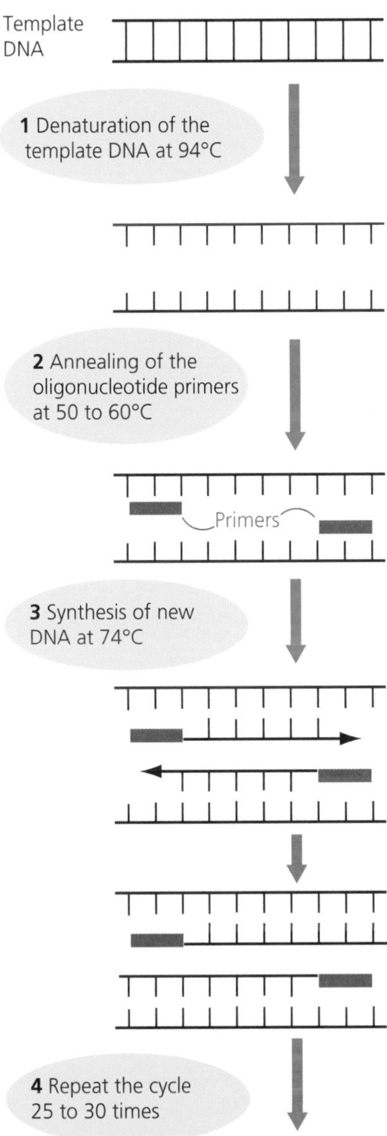

Figure 13.2 The key steps in the PCR. The whole process is done in a machine called a thermal cycler, which can go through the cycle in 3 minutes. 25 cycles take just over an hour, by which time a million copies will be made

Table 13.2 Advantages and disadvantages of in vivo and in vitro gene cloning

Approach	Advantages	Disadvantages
In vivo cloning (in whole cells)	More accurate — fewer mistakes because the cell has 'proofreading' correcting mechanisms Can copy unknown DNA Can reliably copy large fragments of DNA Can also express the cloned gene and make the protein	Takes time Requires a large sample Requires more purification — DNA needs to be extracted from the cell
In vitro cloning (PCR)	Quick — one cycle takes about 3 minutes Works on minute quantities Works on partially decomposed DNA — from old remains, perhaps Simple purification from solution	Lots of copying mistakes — no 'proofreading' mechanisms Not reliable on large fragments Does not work on unknown DNA because complementary primers are needed Cannot make the protein encoded for by the gene

Test yourself

TESTED

1 Starting with a single piece of DNA, how many pieces would you have after six PCR cycles?

Answer on p. 208

Applying recombinant DNA technology

REVISED

Recombinant DNA technology usually involves transferring genes from one species into another. Recombinant DNA is DNA from two different sources that has been joined together. Examples include putting:

- the human insulin gene into bacteria so they will make human insulin for diabetics
- a gene that makes vitamin A into rice crops, so you get vitamin A-rich Golden Rice
- herbicide resistance genes into crops so that spraying kills the weeds but not the crops

Using the insulin example, the main steps are as follows:

1. Obtain the gene for insulin. It is better to make the gene artificially, because the actual insulin gene from the human genome contains introns, which the bacteria cannot splice out. In order for the gene to be expressed, a promoter and terminator region may need to be added, showing the RNA polymerase enzyme where to begin transcribing, and where to finish.
2. Clone the gene — make lots of copies using the PCR.
3. Put the gene into a vector. This involves getting a plasmid (Figure 13.3) and cutting it with the same restriction enzyme that was used to cut the insulin gene. The complementary sticky ends on the plasmid and the gene are joined with a ligase enzyme.
4. Mix the recombinant plasmids with bacteria and treat the mixture so that some of the bacteria take up the plasmid (you don't need details of how to do this).

5 Find the bacteria that have adopted the new plasmid. (See the next section for details.)
6 Culture the **transgenic** bacteria in a sterile environment (grow them in a large vat with a supply of nutrients and oxygen) so that they multiply and express the gene, producing insulin, which can then be extracted and purified.

Test yourself

TESTED

2 Define the following terms:
(a) recombinant DNA
(b) transgenic
(c) plasmid

Answer on p. 208

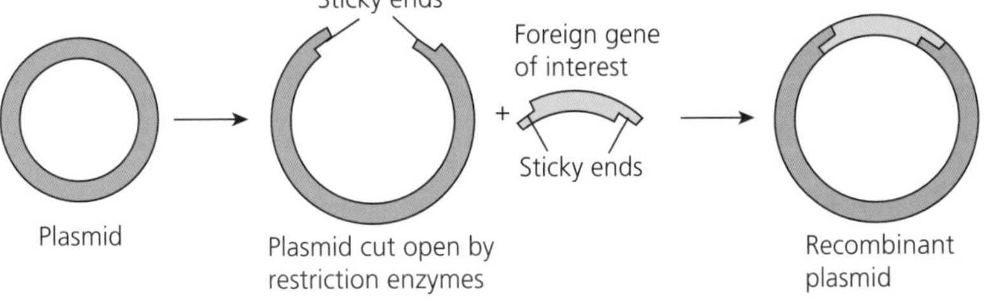

Figure 13.3 Plasmids are useful vectors for 'smuggling' foreign DNA into bacteria

Test yourself

TESTED

3 Why are sticky ends generally more useful to genetic engineers than clean cuts?

Answer on p. 208

Finding the recombinant bacteria

The transformation process is unreliable — most bacteria do not take up the new DNA — so finding the recombinant ones is important. A common method is to add a **marker gene** to the plasmid, alongside the insulin gene. Commonly, a marker gene codes for antibiotic resistance or an enzyme that makes a coloured product. Consequently, only the bacteria that have accepted the gene of interest will survive when grown on a medium that contains the relevant antibiotic, or will show up as a particular colour.

Scientific and ethical objections

There are several objections to DNA technology:

- Genetically modified (GM) crops contain genes that can be transferred to other plant species. This may give rise to pesticide-resistant 'super-weeds' that are difficult to control and may upset the ecosystem.
- The use of viruses as vectors is controversial as it may introduce disease.
- The whole idea of altering an organism's genome by inserting one or more genes is controversial because it may be difficult to predict the long-term effects.

- Cloning whole organisms is unreliable and there are many stillbirths and animals born with defects. It took 200–300 attempts to get Dolly the sheep; when dealing with a sophisticated animal with a nervous system, that amount of suffering is considered by many to be unacceptable.
- **Anti-globalisation** activists are concerned that the large multinational companies will patent GM crops and give farmers little choice about what they grow. For example, there is a large US company that sells GM seeds that are resistant to a certain herbicide, so you can spray it and kill all the weeds but not the crop. It also has a patent on that particular pesticide. So farmers are under pressure to buy both the GM crop and the herbicide.

The AQA specification requires you to 'be able to evaluate the ethical, financial and social issues associated with the use and ownership of recombinant DNA technology in agriculture, in industry and in medicine' and to 'balance the humanitarian aspects of recombinant DNA technology with the opposition from environmentalists and antiglobalisation activists'. Remember that evaluate means 'look at both sides'.

Gene therapy

Genetic diseases cannot be cured. People with **sickle-cell anaemia** or **haemophilia**, for example, have the defective alleles in every cell in their body and that cannot be changed.

However, **gene therapy** brings the promise of treatment by using multiple copies of healthy, working alleles to supplement defective ones. In cystic fibrosis, for example, it may be possible to introduce multiple copies of the working CFTR allele into the lungs. The patient inhales an aerosol — similar to an asthmatic with an inhaler — containing the healthy gene in a vector such as a liposome or virus. The gene is absorbed into the epithelial cells of the lungs, where it is expressed and used to make the missing protein.

This approach is still at the trial stage and few success stories — especially with liposomes — have been reported. One of the problems is how to deliver the alleles to the affected cells, most of which are more difficult to get at than the lining of the lungs. However, gene therapy seems to have great potential for the future.

Exam tip

When asked to state an objection to a particular treatment, such as gene therapy, your answer needs to include some biology. 'We should not play God' is never a good answer. Problems with gene therapy include the possibility that the virus will cause disease or that the long-term effects of altering a cell's genome are difficult to predict: tumours could result, for example.

Test yourself

TESTED

4 Suggest why this aerosol treatment would need to be repeated regularly.

Answer on p. 208

Medical diagnosis

Predisposition to diseases

REVISED

There is an increasing demand for tests that can tell whether individuals possess certain alleles. In addition to well-known genetic diseases such as cystic fibrosis, sickle-cell anaemia, Huntington's disease and haemophilia, we are starting to identify many new alleles that make people **predisposed** to diseases such as heart disease and certain types of cancer.

DNA probes

REVISED

DNA probes are used to find specific base sequences, such as disease-causing alleles. For example, during in vitro fertilisation (IVF), a cell from an embryo can be tested to see if it is carrying one or two alleles for cystic fibrosis. In this way, only healthy embryos can be used, and that particular disease is eradicated from the germline.

The probe is a fragment of DNA that is complementary to part of the base sequence of the allele in question. The probe is added to a tissue or cell sample and binds to (or **hybridises** with) the complementary sequence (Figure 13.4). The fragment is **labelled** with a radioactive or fluorescent marker so that the gene shows up. If the **screening test** is positive, knowledge that an individual possesses a particular gene gives them the opportunity to:

- make certain lifestyle changes — if you find you are predisposed to high blood cholesterol, for example, you can control the fat intake of your diet
- make sure they have all the appropriate tests such as mammograms and prostate gland tests. If the individual concerned is planning a family, their partner may also need to be tested to estimate the chances of the child inheriting the condition. Unborn babies can be tested too.

However, the results of these tests can cause problems: the knowledge that a foetus has a serious medical condition can lead to a difficult decision about whether to continue with the pregnancy, whereas the knowledge that an individual has the allele for Huntington's disease brings the likelihood of an early death. **Genetic counsellors** are there to give advice and support.

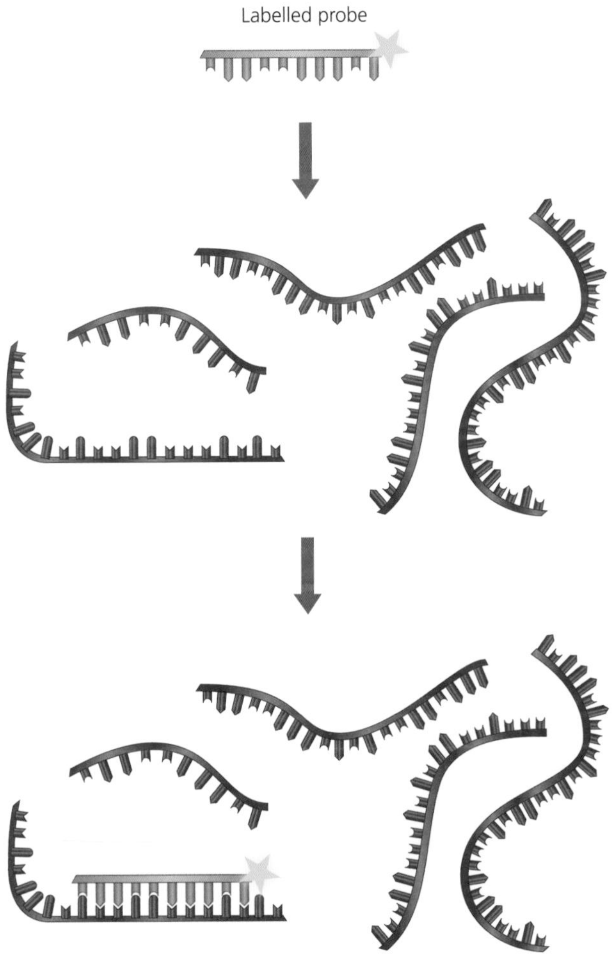

Predisposed is a term frequently used in medicine. It means that the possession of certain genes/alleles makes an individual *more likely* to develop a particular disease.

Labelled probe

Figure 13.4 A DNA probe hybridising with genomic DNA

Test yourself

TESTED

5 What is a DNA probe and what is it used for?

Answer on p. 208

Genetic fingerprinting

DNA profiling

REVISED

The genes of different individuals are remarkably similar — they must be in order to make vital proteins — but the non-coding DNA between the genes varies greatly and is unique to each individual. Specifically, there are many repetitive, non-coding base sequences known as **VNTRs** (variable number tandem repeats). Different people have different numbers of VNTRs, which means that the distances between the recognition sites of a particular restriction enzyme will be different. When digested with this restriction enzyme, the DNA of each individual will have a unique mixture of differently sized fragments. When these fragments are separated by electrophoresis, the result is the familiar banding pattern seen in Figure 13.5. Thanks to the PCR, a DNA profile can be carried out on the smallest of samples, such as a speck of blood or one hair follicle.

Modern DNA profiling often focuses on **short tandem repeats (STRs)**, which are less likely to degrade over time and therefore give more reliable results with older DNA samples.

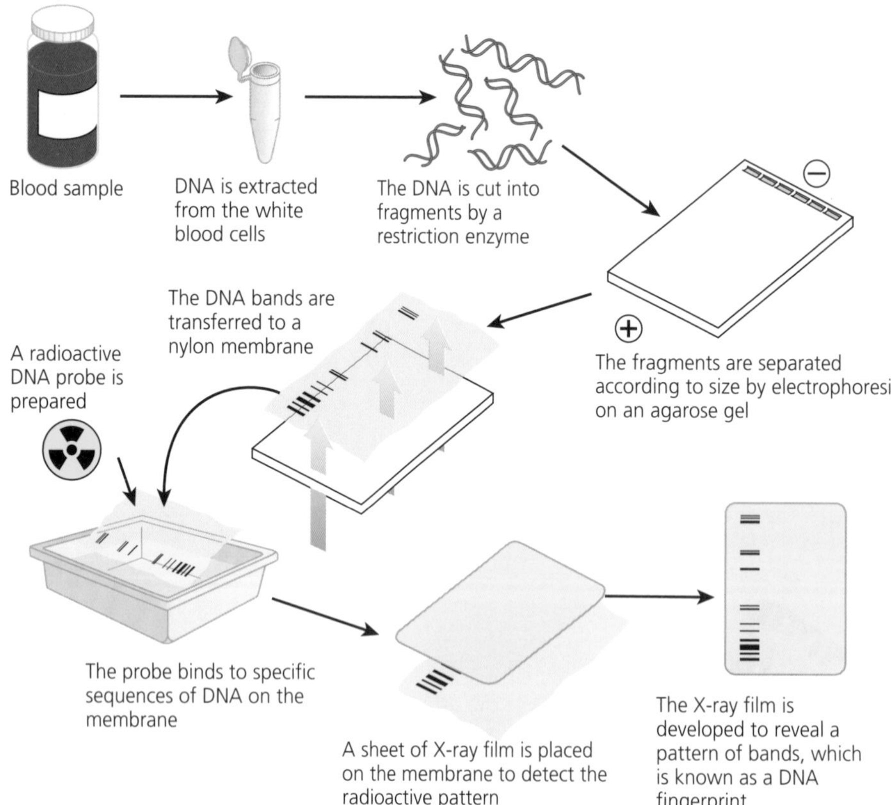
Blood sample
DNA is extracted from the white blood cells
The DNA is cut into fragments by a restriction enzyme
The fragments are separated according to size by electrophoresis on an agarose gel
The DNA bands are transferred to a nylon membrane
A radioactive DNA probe is prepared
The probe binds to specific sequences of DNA on the membrane
A sheet of X-ray film is placed on the membrane to detect the radioactive pattern
The X-ray film is developed to reveal a pattern of bands, which is known as a DNA fingerprint

Figure 13.5 The basic stages of DNA profiling

Genetic fingerprinting can be used to:

- determine **genetic relationships**, such as paternity testing
- determine the **genetic variability** in a population
- ensure **outbreeding** — endangered species need as large a gene pool as possible, so DNA profiling ensures that closely related individuals do not breed together
- establish **pedigree** in the case of thoroughbred racehorses
- identify crime suspects from forensic samples

Exam practice

1 The following diagram shows one type of plasmid used to insert a useful human gene (the gene of interest) into a bacterium.

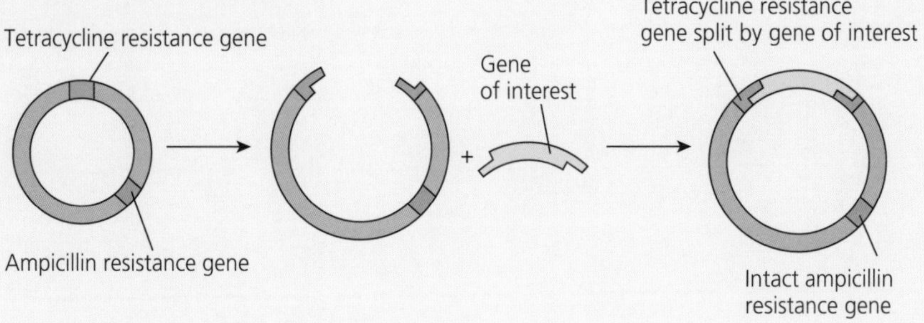

(a) Explain why the gene and the plasmid have to be cut with the same restriction enzyme. [1]

(b) When the enzyme ligase is added to the cut plasmids and the genes, two things can happen:
- the two sticky ends of the plasmids can simply rejoin, or
- the gene of interest can be inserted into the plasmid

Most bacteria do not take up the plasmid at all. The aim of the process is to find out which bacteria have taken up the plasmids with the human gene. Explain how scientists can use antibiotics to isolate the bacteria that contain the gene of interest. [3]

2 The following diagram shows the DNA profiles of a family with four children. The profiles are created by digesting an individual's DNA and separating the fragments by electrophoresis.

Mother **Father** **Daughter 1** **Daughter 2** **Son 1** **Son 2**

(a) Explain why DNA fragments move towards the positive terminal. [2]

(b) Explain why some fragments move faster than others. [2]

(c) Explain how the fragments are made visible. [1]

(d) We inherit our DNA from our parents and any fragments we do not inherit from one parent must come from the other. Use the profiles to determine the biological relationship of each child to their parents. [4]

Answers and quick quiz 13 online

ONLINE

Summary

By the end of this chapter you should be able to understand:

- The three basic ways to obtain a fragment of DNA: cut it out of the genome using a restriction enzyme; make cDNA by working backwards from mRNA; make cDNA by working backwards from the protein. The last two use a 'gene machine'.
- The polymerase chain reaction (PCR) process: denaturing, annealing and extending.
- The process of using recombinant DNA technology to produce transformed organisms that benefit humans. The classic example is using bacteria to make human insulin.
- The use of gene therapy to supply working genes, in order to treat conditions caused by defective genes.
- The use of DNA probes in genetic screening. The use of this information in genetic counselling, in family planning or medical treatment.
- The process and uses of DNA profiling ('genetic fingerprinting').

Now test yourself answers

Chapter 1

1 Condensation

2 Hydrolysis

3 Consists of long, straight, unbranched chains of β-glucose; which lie parallel; and form H bonds along their whole length; forming fibres of great strength.

4 Qualitative tests tell you *what* is present whereas quantitative tests tell you *how much* is present. (Think: quantity = amount)

5 Add ethanol to a sample of cake, shake and then pour off the liquid into water. If the sample contains lipid, a white emulsion is formed.

6 (a) C, H, O

(b) C, H, O

(c) C, H, O, N

7 Proteins, starch and glycogen

8 Enzymes provide an alternative reaction pathway; lower the activation energy; make it easier to achieve the transition state.

9 The transition state is the point at the top of the curve. Once reached, the reaction will go on to completion.

10 Enzymes have an active site that is complementary to the substrate in terms of shape and chemical charges. Different substrates are not complementary.

11 The lock and key model is when the substrate fits exactly into the active site. The induced-fit model is when the active site modifies or changes shape to fit around the substrate.

12 The tertiary structure changes because the weak bonds that maintain it are disrupted/changed/broken.

13 TTGATCCAT

14 Original strand of DNA, nucleotides, DNA helicase, DNA polymerase

15 Hydrogen bonds

16 The nucleus

17 When DNA replicates, in each new DNA molecule one strand is original and one is new.

Chapter 2

1 The roots do not receive light and so they cannot photosynthesise.

2 (a) Eukaryotic

(b) Prokaryotic

(c) Eukaryotic

(d) Prokaryotic

3

	TEM	SEM
Image	2D	3D
Colour	Usually black and white	Can be coloured
Thickness of sample	Must be a thin section	Can be a solid specimen
Resolution	Higher	Lower

4 Four different nucleotides (sugar–phosphate plus A, T, C or G).

5 By facilitated diffusion. (It is water soluble, so it cannot pass through the phospholipids. Therefore, it has to pass through a specific protein channel, which by definition is facilitated diffusion.)

6 Molecules have more kinetic energy, so they bounce around, collide and spread more quickly.

7 Water will pass to the cell with the water potential of −250 kPa because it always passes to the area with the lowest water potential.

8 Plant cells have a cell wall that prevents them swelling and bursting.

9 (a) 50 (because it only goes to equilibrium)

(b) All 100

10 (a) Respiration

(b) Mitochondria

11 A process in which two substances are absorbed together, such as sodium and glucose.

12 Any four from: skin, mucous membranes, stomach acid, blood clotting, lysozyme in tears and sweat, ear wax, low vaginal pH.

13 No. Glucose is present in all cells, so it cannot be used to identify cells as 'foreign'.

14 Antigens are molecules that are not normally found in the body of the host and they stimulate the production of complimentary antibodies. Pathogens and other foreign cells are covered in antigens. Antibodies are proteins produced by B lymphocytes in response to a particular antigen.
15 Antibodies are proteins, and as such are too large and complex to be made in the lab.
16 Breast milk contains antibodies whereas formula milk does not.
17 A type of lymphocyte that lasts for many years and, on exposure to the antigen, can launch a rapid immune response. Memory cells rapidly divide into a clone of plasma calls that can quickly make antibodies.
18 The primary immune response results from first exposure to an antigen. The secondary immune response results from the second exposure to the same antigen. The secondary immune response produces antibodies more quickly, in greater amounts and for longer than the primary immune response.
19 If people are not vaccinated, the percentage of the population that is vaccinated may be too low to prevent the chain of transmission. Therefore, the herd effect may not be effective.

Chapter 3

1 Volume of oxygen used per gram per unit time, for example $cm^3 g^{-1} min^{-1}$.
2 They have a large surface area to volume ratio and can exchange gas over their whole body surface. The diffusion pathways are very small.
3 To ventilate: to pump fresh air into its tracheal system, so as to maintain a diffusion gradient.
4 It is a countercurrent system. It maintains a diffusion gradient along the whole length of the lamellae.
5 An equilibrium would be reached. Half of the available oxygen would be lost.
6 Respiration is the chemical process that releases the energy from organic molecules. Breathing is ventilation: the act of forcing fresh air (or water) over the gas exchange surfaces. (All cells in all organisms respire, but not all organisms breathe.)
7 Many alveoli provide a large surface area. Thin alveolar cells (squamous epithelium) provide a short diffusion pathway. An efficient blood flow and ventilation maintains the diffusion gradient.
8 Gas exchange is slower/less efficient because the alveolar walls have less surface area and are thicker. Activity requires more oxygen for muscular contraction.
9 Condensation reactions involve smaller molecules joining to make larger ones; water is produced. Hydrolysis reactions involve splitting larger molecules to form smaller ones; water is used.
10 In the cell-surface membranes of the gut epithelial cells (they are in the microvilli).
11 The small intestine (ileum).
12 They are not lost in the faeces. They keep on being re-used without having to be re-synthesised, which would be a waste of energy and resources.
13 Bile salts emulsify lipids, making smaller droplets that have a larger surface area for the lipase enzymes to work on and speeding up digestion.
14 The absorbed substances lower the water potential of the blood, so water flows into the blood by osmosis (water follows the solute).
15 There is more room for haemoglobin, so the cells can carry more oxygen.
16 To speed up the delivery of oxygen. (The enzyme speeds up the reaction between carbon dioxide and water, and in turn the resulting acid lowers the affinity of haemoglobin for oxygen.)
17 The affinity of haemoglobin for oxygen varies according to the conditions. The affinity is high in the lungs and lower in the respiring tissues.
18 (a) Coronary artery
(b) Pulmonary artery
(c) Renal artery
19 The cells are thin and permeable, providing a short diffusion pathway and maximising the efficient exchange between tissue fluid and blood.
20 Any three from: oxygen; glucose; amino acids; fatty acids; water; various ions such as sodium, chloride and potassium; vitamins.
21 D, C, B, A
22 No. The atrioventricular valves must shut in order for ventricular pressure to build. If the pressure does not build, the semilunar valves will not open.
23 The atria
24 Left ventricle
25 (a) B and C
(b) It would show the same pattern, but at lower pressure.
26 The stimulus for the heart beat is generated by the heart muscle itself. (If you cut the nerves to the heart, is keeps on beating.)
27 To allow the ventricles to fill with blood before they contract.

28 **cardiac output = stroke volume × heart rate**

= 100 × 160

= 16 000 cm^3, or 16 litres

29 Yes. Someone with a BMI of over 30 is classed as obese.

$$\mathrm{BMI} = \frac{\text{mass (kg)}}{\text{height}^2\ (\text{m}^2)} = \frac{94}{1.56^2} = 38.6$$

30 Transpiration is the loss of water vapour (by evaporation) from the upper surfaces of a plant. The transpiration stream is the movement of water and minerals through the plant in the xylem tissue.

31 Translocation is the movement of organic molecules (mainly sucrose) around the plant form source to sink in the phloem fibres. Transpiration is the loss of water vapour (by evaporation) from the upper surfaces of a plant.

Chapter 4

1 In a prokaryotic cell, the DNA is circular. There is one main chromosome and many smaller loops called plasmids.

2 In a eukaryotic cell, the DNA is linear. Most of the DNA is enclosed in the nucleus. There is some (circular) DNA in the mitochondria.

3 They contain uracil (U) and not thymine (T).

4

DNA sequence	AAT	**CAT**	GTC
mRNA sequence	**UUA**	GUA	**CAG**
Amino acid sequence	**Asn**	**His**	**Val**

5 The tertiary structure is the overall 3D shape of the polypeptide. The quaternary structure is the overall shape of proteins that contain more than one polypeptide.

6 mRNA is variable in length, not folded and contains no hydrogen bonds. tRNA is fixed length, folded and has H bonds.

7 There are 64 types because there are 64 different anticodons.

8 Pre-mRNA contains introns. Mature RNA has had the introns removed (spliced out).

9 62 × 3 = 186

10 B, F, E, D, C, G, A

11 2^{23} = 8 388 608

12 It tends to select for similar individuals with similar genotypes.

13 A group of organisms with observable similarities that can interbreed to produce fertile offspring.

14 Orangutan, gorilla, chimpanzee, human

15 Genus = *Felis*, species = *catus*

16 The number of species present and the number of individuals of each species.

17 Any four from: deforestation, monoculture, removal of hedgerows, use of pesticides, use of fertilisers, other examples of habitat destruction (e.g. construction works).

18 There are fewer tree species and fewer niches for other organisms.

19 Weeds provide niches for insects and other invertebrates, which in turn support the food chain.

20 To avoid the mean being distorted by extreme individual readings.

Chapter 5

1 Chlorophyll, water, ADP and NADP

2 ATP and NADPH (or reduced NADP)

3 In photolysis, water is split to produce replacement electrons for chlorophyll. Oxygen is the by-product.

4 Glycolysis, the Krebs cycle and the electron transport chain

5 (a) ATP is made in glycolysis and the Krebs cycle.

(b) ATP is made in the electron transport chain, powered by hydrogen ions.

6 The folded inner membrane (cristae) provides a large surface area for the reactions of the electron transport chain.

7 (a) 2

(b) 38

8 It is not reduced. It is reduced when it gains an electron, becoming NADH.

9 Volume of oxygen per unit mass per unit time, such as $cm^3\,min^{-1}$ or $g^{-1}\,min^{-1}$

Chapter 6

1 It allows comparisons, for example between different ecosystems, different years or different farming methods.

2 The animal doesn't have to respire as quickly to replace lost heat.

3 Atmospheric nitrogen has a triple bond that is difficult to break.

4 Turning nitrogen gas into ammonium; returning/adding it to the cycle.

5 By nitrogen-fixing bacteria or electrical storms

6 (a) The plant gets nitrate, so it can grow in nutrient-poor soil.
(b) The bacteria are less likely to get eaten and they receive a supply of carbohydrate/sugar made by the plant.
7 The crop will grow without needing as much fertiliser, making it cheaper and leading to less environmental damage.
8 An increased supply of ions leads to increased growth of algae.

Chapter 7

1 Any five from: sound, light, heavy and light pressure, various chemicals (taste and smell), heat, cold, gravity, movement
2 A plant growth response, brought about by cell division and elongation/enlargement.
3 Taxis is a directional response to a stimulus in organisms that can move. Kinesis is a non-directional response to a stimulus in organisms that can move.
4 In a taxis, the organism would go straight to or from the light. In a kinesis, the organism would set off randomly in all directions, but slow down or stop when at its preferred side.
5 (a) They move towards the light.
(b) There is more light for photosynthesis.
6 To avoid predators, to avoid drying out and to increase the chance of finding rotting vegetation (food).
7 To prevent the brain being bombarded with too much information or to allow the body to focus on a new stimuli.
8 Visual acuity is the ability to see detail.
9 The stimulus to beat comes from the heart muscle itself.
10 To give the ventricles a chance to fill.

Chapter 8

1 The cell body
2 Axons take impulses away from the cell body, whereas dendrons take impulses towards the cell body.
3 Active transport and unequal facilitated diffusion
4 –70 mV
5 The action potential arrives at the synapse → Calcium ions flow into the presynaptic membrane → Molecules of neurotransmitter diffuse across the synaptic cleft and fit into specific receptor proteins → The permeability of the postsynaptic membrane changes → sodium ions flow in, causing a positive charge to build up inside the postsynaptic membrane → If the threshold is reached, an action potential is created in the postsynaptic neurone
6 ATP is needed for movement of vesicles, resynthesis of the neurotransmitter and active transport of calcium ions out of the synaptic knob.
7 If they were not, the result would be chaos and there would be no coordination. Impulses could pass to sensory organs down sensory nerves and back from muscles down motor nerves.
8 The effects, in order, are inhibit, inhibit, inhibit, prolong, prolong.
9

Component	Role in contraction
Actin	One of the main proteins — the thin filaments
Myosin	The other major structural protein — the thick filaments that have many moveable heads
Troponin	Small, globular protein bind to troponin, moving it aside and exposing the myosin binding site on the actin
Tropomyosin	Long, thin fibrous protein that blocks the myosin binding site on the actin
ATP	Binds to the myosin head; splits to provide the energy to detach the myosin head and reattach further along
Calcium	Initiates contraction by activating tropomyosin

Chapter 9

1 A positive feedback is a mechanism for change ('change creates more change'). A negative feedback is a mechanism for stability. A change brings about a mechanism that reverses the change.
2 (a) Proteins
(b) Amino acids, glucose
(c) Urea
(d) Water, sodium ions, potassium ions and chloride ions
3 Because there is always metabolic waste (urea) that needs removing.
4 Both hormones:
- travel in the blood
- fit into receptor proteins in the cell membrane
- do not enter the cell
- alter the activity within the cell via adenyl cyclase and cyclic AMP
- activate specific enzymes within the cell

5 A change in water potential is detected (by the hypothalamus). A corrective mechanism is initiated (ADH causes more water to be reabsorbed). The mechanism is switched off when levels return to normal.

Chapter 10

1 An alternative form of a gene

2 Zero

3 Alleles that, if present, will both be expressed.

4 Red, pink, white in the ratio 1:2:1

5 A or B

6 Individuals 9, 11 and 14 are all born to non-albino parents. These parents must be carrying the allele without expressing it, which is the definition of a recessive allele.

7 A gene carried on a sex chromosome

8 Females have two copies of each allele; males have only one copy. Therefore, in males, each X-linked allele is expressed because there cannot be a second/dominant allele to mask its effect.

9 Yes, but the mother must be a carrier and the father must be a haemophiliac, which is very rare.

10 See table

Individual	Genotype	Reason
1	X^bY	He is a colour-blind male
2	X^BX^b or X^BX^B	She is a normal female and she gave X^B to individuals 4, 5, 6 and 8; she may or may not have had a second X^B
3	X^bY	He a colour-blind male
4	X^BX^b	She has a colour-blind daughter (individual 11), so she must be carrying X^b
5	X^BY	He is a normal male
6	X^BX^b	She has normal vision, but has a colour-blind son (individual 12); individual 7 has normal vision, so the b allele must come from her
7	X^BY	He is a normal male
8	X^BY	He is a normal male
9	X^BX^b	She has normal vision, but has a colour-blind son (individual 14). Individual 8 has normal vision, so the b allele must come from her
10	X^BX^b	She has normal vision but must have inherited a b allele from her colour-blind father
11	X^bX^b	She is a colour-blind female (quite rare)
12	X^bY	He is a colour-blind male
13	X^BX^B or X^BX^b	She is a normal female, but there is no way of telling what the second allele is
14	X^bY	He is a colour blind male
15	X^BY	He is a normal male

11 (a) Brown, short: brown, long; albino, short; albino, long. In equal numbers (or 1:1:1:1)

(b) 3 Brown, short to 1 brown, long. (No albinos.)

12 (a) Genes whose loci are on the same chromosome.

(b) Independent segregation separates whole chromosomes, so linked genes are always inherited together. Crossing over is the only process that can separate linked genes.

(c) There are many more genotypes like the parents' genotypes, suggesting linkage.

13 (a) Yellow flowers result from **AAbb**, **Aabb**.

(b) White flowers result from **aabb**, **aaBB**, **aaBb**.

14 They add up to more than 1.

15 All the different alleles circulating in a population.

Chapter 11

1 Population means all of the individuals of one species. Community means all the individuals of all species.

2 Biotic factors: predation (e.g. leaf-eating insects), disease, competition from other trees. Abiotic factors: any three from temperature, soil moisture, soil mineral ions, light intensity.

3 (a) Quadrats (small ones, say 25 cm^2)

(b) Belt transect

(c) Interrupted belt transect

4 $\frac{200 \times 200}{23}$ = 1739 (rounded down)

5 When a barren habitat becomes inhabited by pioneer species.

6 Existing species make conditions more favourable.

7 The final stage of succession: a stable community.

8 A mixture of rotting organic matter and bacteria that releases mineral ions slowly.

9 The rate of increase in biomass (weight of all the organic molecules such as starch, cellulose etc.).

net productivity = gross productivity – respiration

10 Units of energy or mass, per unit area, per unit time — for example, kg m^2 day^{-1}

11 At first there are relatively few species and individuals (just the pioneer species), so not much photosynthesis occurs.

As succession occurs, there is greater diversity and more individuals, so more productivity.

When the ecosystem stabilises (climax community) there tends to be a lot of trees, diversity is lower again and the trees prevent most photosynthesis in smaller plants below.

Chapter 12

1 The mutation might alter the amino acid sequence but leave the active site unchanged.

2 Males make more gametes than do females.

3 A clone is a genetically identical copy. This can be a piece of DNA, a cell or a whole organism.

4 The primary structure (amino acid sequence) will be altered and so the tertiary structure will be different. The protein will probably be non-functional.

5 Red blood cells do not contain DNA at all. Eggs (ova) and sperm contain one copy of the genome. The X and Y chromosome factor complicates the issue even further.

Chapter 13

1 64

2 (a) DNA from two different species joined together.

(b) An organism containing recombinant DNA.

(c) A circular piece of DNA found in bacteria.

3 Sticky ends are staggered cuts in the DNA. They can be joined to complementary cuts made by the same enzyme.

4 Epithelial cells have a high turnover — they are lost and replaced constantly.

5 A short piece of labelled DNA that is complementary to a target sequence. It is used to find a particular gene/allele/base sequence.